A TILTROTOR ENTERPRISE

From Iraq to the Future

ROBBIN LAIRD

SLDinfo.com

In recognition of the contribution of LtGen (Retired) Fred McCorkle to launching the tiltrotor enterprise.

Without Lieutenant General (Retired) Fred McCorkle there would be no V-22 Osprey today. Seldom does one leader standout among the many important contributors it takes to develop an innovative and highly successful defense program, but in the case of the Osprey, Fred McCorkle is a giant among men.

As an aviation requirements officer serving in Headquarters Marine Corps during the 1980's, McCorkle was there at the birth of the tiltrotor concept. Later, as the Corps' Deputy Commandant for Aviation from 1998-2001, he stemmed the tide against those who wanted to abandon the program in the aftermath of two tragic accidents that occurred during operational testing.

Rather than walk away from the cornerstone of the Marine Corps' vision for projecting naval power ashore, McCorkle emphasized the V-22's superior capabilities compared to traditional helicopters. He advocated technical solutions and tactics, techniques and procedures that ultimately led to full operational capability and a series of successful combat deployments into Iraq and Afghanistan beginning in 2007.

This book is dedicated to the perseverance and foresight it took to save the Osprey for future generations of visionary leaders. Today, we are the beneficiaries of General Fred McCorkle's commitment to ensuring his fellow warriors would have the finest technology available to fight and win our nation's future battles.

PROLOGUE

"I sort of think of it like a game of chess. I think of a traditional or legacy ARG-MEU as being able to move a pawn one space at a time towards the enemy.

"If you have ever played chess it sometimes take a while to engage your opponent. We now have the ability to move a knight, bishop, or rook off of this same chess board and attack 180 degrees towards the rear of our enemy. We can go directly after the king.

"Yes, it's not really fair, but I like that fact. The speed, range, and don't forget the reliability of the MV-22 allows me to do this.

"We talk about staying ahead of the bow wave. Well there is a tsunami of change coming when we talk about the ability to fight an enemy and to support Marines ashore.

"We can increase our area of operations (AOR) exponentially because we can spread out our ships; now we have an aviation connector in the Osprey that can move Marines a tremendous amount of distance and in a very short amount of time."

. LtCol Chris Boniface, ACE Commander of the 26th MEU, 2012 interview.

FOREWORD

When Robbin Laird first approached me and let me know he was planning to write a book about the Osprey, an aircraft that has been and continues to be controversial, I was pleased that an author of his caliber was taking up the task.

As is Robbin's modus operandi, he has approached this book by relying on numerous individual accounts of the people who were there, those who actually lived and experienced firsthand, the history of this remarkable aircraft.

When I *"replay the tape"* of my personal history with this aircraft, spanning a 37-year career of operational, staff, and combat assignments, there are a few themes and events that stand out in my memory.

On the morning of April 1, 2003, I led the first combat mission for Marine Medium Helicopter Squadron 263 (HMM-263) Reinforced, the Aviation Combat Element (ACE) of the 24th Marine Expeditionary Unit (Special Operations Capable 24MEU(SOC)) in support of the early stages of Operation Iraqi Freedom (OIF). The flight consisted of two CH-46E Assault Support aircraft, and two AH-1W Cobra Attack Helicopters.

We launched from the deck of the Amphibious Warship USS Nassau, LHA-4, and we were approximately 30 miles to sea. The mission was to get to Jalibah Airfield and to conduct liaison with Marine Aircraft Group 29 (MAG-29) to facilitate the inflow of the remainder of the HMM-263 Rein ACE in order to support follow on combat operations.

It was roughly 30 nautical miles to sea when we launched, and we knew the fuel would be tight, but we had planned accordingly. Bottom line, during ingress we had little room to maneuver if required to accommodate the airspace concerns of other warships, and the Northern Arabian Gulf looked like a warship parking lot, and we knew we had about ten minutes of additional fuel if it became necessary to address a ground threat.

We made it to Jalibah with less reserve fuel than planned after having to alter our inbound course due to the rogue Iraqi forces setting fire to the oil wells. This portion of the mission was successfully completed but getting back to the USS Nassau that evening is another story, for another time.

Fast forward approximately five years to July 22, 2008 and I was the Commanding Officer (CO) of Marine Medium Tiltrotor (in the Marine Corps tiltrotor is always one word by the way!) Squadron 162 (VMM-162), the Golden Eagles, part of MAG-16, 3d Marine Aircraft Wing Forward.

Senator Obama, who was running for president, along with Senators Jack Reed and Chuck Hagel, visited the troops in Iraq. Our mission was to get them out of Iraq, to Amman-Marka International Airport in Jordan for the rest of their international trip. Very long story short, we launched five aircraft out of Al Asad Airbase, including a couple of backups, flew east to a small, single aircraft landing zone in Al Ramadi, which required the other aircraft to loiter at altitude waiting their turn. We executed the pickup and the flight flew west at altitude, out of the danger of small, medium, and heavy weapons as well as RPGs, crossed the border into Jordan and landed at Amman-Marka International Airport.

Needless to say, the congestion on the airport was almost

overwhelming, and we ended up spending about 45 minutes on the ground, but never shutdown and did not get fuel.

After successfully delivering our distinguished visitors, we departed Jordan, crossed back into Iraq and realized we had more fuel than we had planned and we were being requested by the Direct Air Support Center (DASC) to pickup a few Assault Support Requests (ASRs). We happily obliged and each aircraft serviced a couple of ASRs before landing at various Forward Operating Bases (FOBs) and refueling for the flight back to Al Asad.

The mission to Jalibah Airfield spanned approximately 170 nautical miles. We completed the mission within all Naval Air Training and Operating Procedures Standardization (NATOPS) limits, and luckily Murphy's Law didn't show up.

By contrast, the mission to Amman-Marka International Airport covered approximately 800 nautical miles and the flight profiles varied from at altitude (-12,000 ft) to include holding, low altitude ingress and considerable engine runtime on the ground. The difference between my beloved CH-46E, affectionately known as the Battle Phrog, and the MV-22 Osprey couldn't have been more stark in my view.

VMM-162 remained in Iraq until October supporting combat operations and the aircraft performed remarkably, and credit must be given to the amazing Marines and our patriotic civilian maintainers who kept the squadron in the fight. We didn't drop a single mission.

Dr. Laird's book captures what has been nothing less than a technological revolution in aviation. And just so it is completely clear, the Osprey isn't just a new aircraft, tiltrotor flight is a completely new way to fly, a new way to operate, and its impact on the United States Marine Corps has been very consequential, and the journey isn't over.

Robbin Laird clearly understands that and has captured it in this book. The Osprey the Marine Corps is operating today, compared to the aircraft I flew in Iraq as the CO of VMM-162 in 2008, is more capable in terms of augmented flight, software upgrades, survivability, and flight envelope expansion, which included an engine power upgrade.

There have been nine Deputy Commandants for Aviation (DCA) since the Osprey's Initial Operational Capability (IOC) almost 20 years ago in 2007, and under the stewardship of those DCA's and the Aviation Hallway, the aircraft has become more capable and hopefully that there is more capability growth to come.

The Marine Corps will fly the Osprey for at least 50 years, to the year 2060 or most likely well beyond. I hope Robbin's book will contribute to the acknowledgement that the MV-22 has been an absolute success, and it is time, as we approach the 20th anniversary of the aircraft's IOC, that we look to make the investment to ensure this one of kind capability, stays healthy and the community robust.

Over the years, as I've conducted interviews and had discussions with Dr. Laird, it occurred to me, how did an aircraft with the ability to carry our nation's sons and daughters into and out of harm's way, farther, faster, and at lower risk, somehow become maligned as an example of government waste and come under unprecedented, if not absolute irrational scrutiny, despite a safety record that previous rotor craft could only dream of?

Dr. Laird has been interested in my perspective because I had a front row seat to this constant mismatch between rhetoric and reality that has taken place over the last two decades.

Dr. Laird has done an incredible job capturing the reality, the true story of the MV-22 Osprey...an assault support aircraft without comparison, and I highly commend this book.

Lt Gen (Retired) Karsten Heckl most recently served as the Commanding General, Marine Corps Combat Development Command (MCCDC)/Deputy Commandant for Combat Development and Integration (CD&I) for the United States Marine Corps, where he oversaw the intellectual and institutional epicenter for the evolution of the Marine Corps, focused on strategic vision, requirements analysis, innova-

tive concept development, technology strategy and solution-building. He supervised the entire requirements enterprise for the Marine Corps from initiation to fielding.

During LtGen (Ret.) Heckl's more than 36 years in the U. S. Marine Corps, he has commanded at every level, and his assignments have included multiple tours with Headquarters Marine Corps (HQMC), various joint billets and numerous operational and combat assignments. In his role as CG MCCDC/CD&I, he launched new concepts and new technology-driven equipment, such as The Functional Concept for Maritime Reconnaissance and Counter-Reconnaissance, The Concept for Stand-in Forces, and the Remotely Operated Ground Unit for Expeditionary (ROGUE) fires – Navy/Marine Expeditionary Ship Interdiction System (NMESIS).

LtGen Heckl (Ret.) has served as the Commanding General of both the Second Marine Aircraft Wing and the First Marine Expeditionary Force, the largest warfighting formation in the Marine Corps, and he was the Chief of Staff for Naval Striking and Support Forces NATO and he was the Assistant Deputy Commandant for Aviation, HQMC.

LtGen (Ret.) Heckl commissioned through Platoon Leaders Class (PLC) at Georgia State University with a B.B.A. in Accounting, and he has been recognized with numerous personal and unit awards for his service. He also holds an M.A. from the Naval War College in National Security and Strategic Studies.

CONTENTS

INTRODUCTION

As the United States faces a global overload of strategic challenges and the concomitant challenge of shaping an effective and capable force to deal with these challenges but having serious budget stringencies, leveraging the unique capabilities which the United States already possesses is crucial.

It is nice to think of 6th generation aircraft, new AI autonomous systems, new weapons, and the like, but adapting what you have and leveraging unique capabilities which you possess is a key part of the way forward.

Whether it be the Aegis global enterprise or the F-35 global enterprise or the tiltrotor enterprise, the United States has shaped unique warfighting capabilities which it can leverage as it shapes effective forces moving forward for today's and tomorrow's challenges. And by doing so, position American forces to incorporate or adapt to new autonomous capabilities as well.

I have written extensively about the Aegis global enterprise and the F-35 global enterprise but we should also focus on a core capability which the United States has crafted and evolved since its introduction into Iraq in 2007, namely the tiltrotor enterprise.

If the Chinese had developed this capability and had built on its use since 2007 and its proliferation in the joint force, I guarantee there would be a robust literature on this threat and how do deal with it.

But since we have done it, we spend as much time criticizing it as understanding how the tiltrotor enterprise has transformed the capabilities of the USMC, the USAF and now the U.S. Navy with the U.S. Army next up.

This is a story of a unique capability which has reshaped the USMC in ways that are unimaginable without it. It has given the USAF special operational capabilities and now the U.S. Navy will experience a very different capability and approach to sustaining its distributed fleet in the face of contested logistics.

And as the U.S. Army focuses on how to distribute its force, the new tiltrotor capability will become a backbone for an effort to leverage speed and range which no rotorcraft possesses.

Over the years, I have done many interviews with USMC operators and maintainers of the MV-22B. And with the development of a Navy variant, the CV-22B, I have begun to interview those involved in shaping this capability for the Navy as it crafts its new approach to distributed maritime operations. I briefly introduce the new Army variant but interviews are just starting with regard to the new variant, which will be a focus of future work.

The timeline of the operational development since the introduction of the MV-22 in 2007 in Iraq has seen the expansion of the concept of operations of the USMC as the aircraft numbers and use multiplied over the past decade and a half.

The learning curve of the USMC and the evolution of industrial support and engineering capabilities for the platform have shaped new ways to use the aircraft for distributed operations across the spectrum of warfare.

U.S. Marines with Marine Aircraft Group (MAG) 26 prepare to fly MV-22B Ospreys at Marine Corps Air Station New River, North Carolina, Aug. 9, 2022. U.S. Marine Corps photo by Lance Cpl. Christian Cortez.

And the creation of a core industrial capability to shape the drive forward in tiltrotor evolution coupled with the innovations of the Marines, the Air Force and the Navy in using the aircraft have created a unique tiltrotor enterprise.

- How did we get here?
- And what is the path forward?
- And how might the U.S. military leverage this unique capability moving forward to deal with strategic challenges they face in global operations?
- And how has the payload revolution which has enabled a kill web transformed the Osprey into a multi-domain warfighting capability as well?[1]

And in answering these questions, I would note that this book could not have been written except for the time given to me by Marines at all levels within the USMC. I would thank Navy officers and officials as well. The analysis in this book has been based on access and observation. It is not an armchair book at all.

I wrote a piece in 2015 which highlighted the importance of such an approach.

I have visited 2nd MAW many times over the past few years and have found the leadership and warriors of the 2nd MAW to have been engaged in global operations and while doing so finding ways to reshape the capabilities of the USMC to execute 21st century missions.

2nd MAW was an additional important stimulant to my own thinking.

After taking several journalists to Cherry Point, and having a chance to meet the senior leadership and several Marines, not a single story emerged. This was in spite of the 2nd MAW team being on the cutting edge of introducing the Osprey into combat, about to go on their final tour into Iraq, and to have discussed with passion why the massive buy of MRAPS made no sense.

In fact, the CG asked the relevant LtCol to discuss why he thought the MRAPs were not good things to hand over to the Iraqis. In giving a spirited answer that history would prove correct – to the point of they were not going to be maintained by the Iraqis and that there were too many variants to be a useful leave behind, among other points, I thought that a story like that in 2007 would make good sense.

But it did not see the light of day.

Why?

I asked one prominent journalist why he did not pursue the story.

The answer: "The Marine who made these points was only a Lt Col. When I deal with the U.S. Army on such matters I only deal with Generals."

OK, there clearly was a problem here.

And looking around, I found that indeed the apparently least inter-

esting thing to discuss is what the warriors were actually doing in shaping their combat futures.

And it was at this time of course, all the cubical commandos and asserted facts journalists buried the Osprey into the proverbial sand of history.

Next up of course has been the F-35.

As one prominent analyst told me recently: "I do not know why the USMC is taking risks with the plane, for they should wait until the GAO and the testers are done."

Of course, one answer is that the GAO and the testers are never done, and new equipment is not operated by either group.

Another answer of course is that the GAO is better at blocking acquisition than aiding it.[2]

In the book which Ed Timperlake and I have done on MAWTS-1, we highlighted the USMC approach — give it to the operators and push the envelope on operational capability.[3]

This is why the Marines have been the driver of the tiltrotor enterprise and why it now is a multi-mission capability crucial to the strategic transition from the land wars to shaping an effective distributed warfighting force.

But the USAF and the U.S. Navy have been key players as well. And now the U.S. Army is designing a new more sustainable variant of the Osprey which will expand significantly the numbers of operational tiltrotors and thereby expand their operational impact on the joint and allied forces.

In this book, I tell the story of the evolution of the tiltrotor aircraft from the time of its introduction into combat in Iraq in 2007 and begin the story of the development of the new variant of the aircraft being designed by Bell and the U.S. Army.

This book is the first of two volumes written to highlight the tiltrotor enterprise story, This book crafts the core narrative and includes some of the interviews I have done along the path of my journey with the tiltrotor aircraft. But I simply have done too many interviews to include in one volume.

So I have written a companion volume which includes additional interviews which I have done over the years which provide further details on the journey. In that second volume to be published in the Fall of 2025, I have included essays as well by other analysts to fill out an understanding of the evolution of the tiltrotor enterprise.

That volume is entitled: *The Tiltrotor Perspective: Exploring the Experience*.

Osprey flying overhead near my house in OBX.

And finally, a personal aside.

Out over the water in the bay on which my house in the Outer Banks is located sits an Osprey nest.

I have learned much from watching them fly, hunt and fish. Most importantly, you don't want to be on the receiving end of an Osprey insert!

1

LOOKING BACK AND LOOKING
FORWARD WITH THE OSPREY

I n August 2024, LtGen Heckl retired from the USMC after 37 years of service. He has held several command positions and has a wealth of combat experience. I conducted a series of interviews with him in which we discussed both his experience and his judgement with regard to the way ahead for the joint force as it addresses the challenges of the evolving strategic environment.

In the first interview in that series, published on September 16, 2024, we focused on his experience with the Osprey and how he views its role and impact on USMC operations. We discussed this from the standpoint of his own experience with the aircraft beginning with his initial engagement with the program, his deployment to Iraq and then his experience with the aircraft as it has evolved over time within the context of USMC operations.

This interview provides a good place to start when discussing the tiltrotor enterprise.

We started with his initial involvement with the aircraft. He came to the Osprey from a CH-46 background and operations with that aircraft in a variety of combat situations. Heckl noted that when he left as a MAWTS-1 instructor in 1999, he had been selected as one of the initial members of the new VMMT-204 squadron to replace HMT-204. There were two accidents involving the aircraft in 2000 which slowed down the process of launching VMMT-204 so Heckl left to deploy with HMM-263 which formed the Aviation Combat Element for the 24th Marine Expeditionary Unit and participated in the initial combat operations in Operation Iraqi Freedom.[1]

Heckl then came back to Washington where he was attached to the office of the Deputy Commandant for Aviation, Headquarters Marine Corps, where he worked as the requirements officer for the Osprey. He first flew the aircraft in May 2000 and by the time he was back in Washington he had accumulated 160 hours on the aircraft.

Heckl was now a Major working for the Deputy Commandant for Aviation. And during that time Col Glen Walters, who later became the 34th Assistant Commandant of the Marine Corps from 2016 to 2019, was the Commanding Officer of VMX-22 and putting the Osprey through Operational Evaluation, specifically Operational Test-IIG, testing the platforms operational effectiveness and suitability.

I dealt often with VMX-22 in the past, but it has now become VMX-1.

As one source described VMX-22:

"Marine Tiltrotor Operational Test and Evaluation Squadron Twenty-Two (VMX-22) is a United States Marine Corps tiltrotor squadron consisting of MV-22 Osprey aircraft. The squadron, known as the "Argonauts", is based at Marine Corps Air Station New River, North Carolina. VMX-22 stood up in August 2003 and reports to the Commander, Operational Test and Evaluation Force (COMOPTEVFOR), who in turn reports test data and results to the Office of the Secretary of Defense, Director, Operational Test and Evaluation."[2]

Heckl then got selected for command of an Osprey squadron which deployed as the second squadron into Iraq in 2008. It was a seven-month deployment to Al Assad Airbase in

Iraq and the deployment spanned the brutal summer months and the aircraft according to Heckl "performed spectacularly" primarily a function of the hard work by the maintenance Marines.

I asked him to take us back to that initial period to remember what were the expectations at the time with regard to the initial Osprey deployments.

"We knew that the aircraft had unique speed, range, payload flexibility, unique survivability capabilities.

"But our awareness of what it could do came with the use of the aircraft in real world situations.

Senators Barack Obama (D-IL) and Jack Reed (D-RI), exit an MV-22 Osprey during the congressional delegates visit to al Anbar province, Iraq. Marine Medium Tiltrotor Squadron 162, from Marine Aircraft Group 16, 3rd Marine Aircraft Wing (Forward), transported Obama, Reed, and Sen. Chuck Hagel (R-Neb) to several stops throughout the province. Credit: Marine Corps Air Station Miramar

"For example, Senator Barack Obama came over with Sec Def Hagel and Senator Jack Reed. We had to support Senator Obama moving around. I launched five airplanes. I was in one of them. But we only needed three of them but we launched backups to be ready.

"We had to land at a Forward Operating Base (FOB) just east of the Baghdad International Airport. We had to do a vertical landing in sequence with three aircraft to carry all of the people involved.

"We flew all the way across the country of Iraq, into Jordan, and went to Marka International Airport to drop off Senator Obama. We landed. We didn't shut down, didn't get gas, and we were on the deck for probably 20-30 minutes after dropping him off and we took off again, flew back into Iraq and had enough gas to service two assault support requests before we got gas.

"If we were using CH-46s that would have taken us two days and several landings for gas.

"That was an eye opener for me, and I thought back to when we invaded Iraq back in 2003. Flying the Battle Phrog, I launched off the USS Nassau. I barely made it to Jabala airfield, before we had to get gas. And here, the V 22 was traversing almost the entire width of Iraq, and covering a third of the country of Jordan, and then back into Iraq on one bag of fuel. Just amazing.

"The other part about the airplane that we realized was survivability. So back then, the zones were designated a color code. Green is obviously good. Black was the worst. So if you had a black zone, you could go only go in at night. The Osprey operated in the black zones comfortably.

"We would ingress to the objective while we wore night vision goggles. And then at the pre-brief point, we would pull the thrust control lever, the TCL or the gas pedal, essentially to flight idle and then we would just start the slow, gradual spiral.

"We had very minimal infrared (IR) signature. The most significant contributor to IR signature is hot gas impingement. Hot gas comes out of the 46 or a 53 and heats up the surface of the aircraft.

"With the V-22 its two big heat generators are the engines. Those two powerful engines are located on the wing tips, sitting at the dead center of a 38-foot proprotor that's blowing hurricane force air around it. And both engines have full time infrared suppressors. And not to mention that while you're doing that, your acoustical signature is virtually eliminated, you're just basically gliding on the wing. You got down to a certain altitude, and you'd level off, maintaining at or below 200 feet AGL above ground level.

"So we stay out of the small arms fire and RPG envelope, and reduced our signature to MANPADS. So the heat signature is low, the acoustical signatures is low, and then less than a mile from the landing zone is when you actually have to start putting in power to do a vertical landing.

"In other words, the survivability is just phenomenal. Of course, the systems on the aircraft since then have been upgraded. The automation has progressed by leaps and bounds. There is virtually no LZ the aircraft can not land in regardless of dust condition."

MV-22 Ospreys with Marine Medium Tiltrotor Squadron 162, Marine Aircraft Group 16, 3rd Marine Aircraft Wing land at an airport in Amman, Jordan, July 22, 2008. Four aircraft from the squadron transported Senators Barack Obama (D-IL), Jack Reed (D-RI), and Chuck Hagel (R-Neb) from al-Anbar province, Iraq to Jordan, July 2008. Credit: Marine Corps Air Station Miramar

I pointed out that the discussion of Osprey safety – which is about the same level as most rotorcraft – has completely ignored the impact of the aircraft on survivability. How many lives have been saved by the tiltrotor aircraft and how it operates is not a common subject.

Heckl agreed completely and argued that safety is a factor in survivability, but the focus has to be on survivability.

"As a former MAWTS-1 instructor and then as CO, our focus was very much on the Tactics, Techniques and Procedures (TTPs) for enhanced survivability. And with regard to rotorcraft, a key survivability challenge comes with regard to ingress and egress from a LZ. With the Osprey, we can approach or leave the LZ from a variety of points differently from rotorcraft and with speed they simply do not have. In Iraq, we would go through the small arms and RPG envelope in seconds."

In 2010, Heckl became the CO of MAWTS-1 where the first class of Ospreys were integrated into the TTPs being worked at this unique warfighting center.

Heckl argued that the coming of the Osprey to MAWTS-1 in 2010 was part of honing the warfighting edge of the USMC going ahead. "Professionally, sound, and tactically planned missions are inherently safe. That is what we focused on in MAWTS-1. Safety is part of survivability.

"For a combat pilot, safety is part of the survivability issue. And that's why I love MAWTS-1.

"We now had the V-22 collaborating with all the other rotorcraft and for me V 22 needs to be discussed in terms of operating with the KC-130Js. I would express caution for anybody to discuss V 22 without every other sentence talking about KC-130J. It is a phenomenal workhorse and when paired with the V-22 creates an incredible operational envelope for the Marines."

His next assignments after leaving MAWTS-1 in 2011 and until his becoming the commander of 2nd Marine Air Wing in 2018 included the following: J3 Director of Operations, United States Forces- Afghanistan (USFOR-A), Kabul, Afghanistan, Military Assistant to the Secretary of the Navy, Assistant Deputy Commandant for Aviation, HQMC Aviation Department, Washington DC, and Chief of Staff, Naval Striking and Support Forces NATO (STRIKFORNATO), Lisbon, Portugal.

For the purposes of this interview, we next focused on his time serving as the Military Assistant to the Secretary of the Navy. Heckl related his experience in working with the Sec Nav and introducing him to the Osprey and its impact. And this experience would lead the Navy to start seriously its journey to acquire the aircraft for the replacement of the C-2 in the support role for the large deck carriers.

This is a story I have never heard before and frankly is of great interest to understanding the growth of the tiltrotor enter-prise. At the time during this tour in the Pentagon, the Sec Nav was Ray Mabus and the CNO was Admiral Greenert.

According to Heckl: "I am now a Colonel, and the Sec Nav wants to go out on the Bush to see the X-47 unmanned platform operating off of the large deck carrier. The staff is organizing the visit whereby he flies by aircraft to Norfolk and then take the C-2 aircraft onto the carrier.

"I observe that the Secretary is not pleased with how much time this going to take. I suggested to the Secretary that there was a way to go from the Pentagon helo pad directly to the Bush. I went to meet LtGen Schmidle, DCA, and he offered an Osprey. The Secretary used the aircraft and when he returned to the Pentagon, he turned to the CNO and asked: "Why are we not buying this?"

Heckl went on to work with LtGen Davis as assistant DCA, Commander of 2nd MAW, CG of 1 MEF and retired as Commanding General Marine Corps Combat Development Command. In those positions, he saw the Osprey evolve into a multi-mission aircraft with an ability to perform very flexible roles with its roll on and roll off capabilities. I talked with Heckl when he held all of those positions about aviation and combat development issues and have published these interviews throughout the years.

But we concluded by discussing the growing relevance of the

Osprey to the joint force and the USMC with the focus increasingly being upon distributed operations.

According to Heckl: "Logistics, C2 and maneuver are three key warfighting functions. The Osprey can deliver all three to a distributed force in unique ways. We have only scratched the surface of what this aircraft can deliver for the distributed force. But to get full value our of our fleet, we need to invest in and provide mid-life upgrades."

We then discussed a subject which is not often the focus of defense modernization discussions, which is really the question of paying for the stretched service lives of equipment and in this case aircraft. The Marines used their aircraft extensively in the Middle Eastern land wars, and the investments in repairing and modernizing those aircraft was never really made. The Marine Corps and its air capabilities has been stretched and needs near term investments to sustain its operational capabilities into the future.

The Osprey is a clear case of such an aircraft, one which needs mid-life upgrades to extend its service life. The B-52 is still relevant because of regular investments in upgrades; the Osprey needs to have such a funding path as well.

❧ 2 ❧

COL SPAID LOOKS BACK AT
THE OSPREY EVOLUTION

A nother way to look at the building and evolution of the tiltrotor enterprise is to share the experience of a plank holder in that enterprise or what was called early on the "Osprey nation".

I have had the privilege of interviewing LtCol Spaid in 2014 and 2021 at USMC Air Station New River. The two interviews were seven years apart are interesting to read in sequence which provides insights into the tiltrotor journey. He was LtCol Spaid in 2014 and Col Spaid in 2021.

THE 2014 INTERVIEW

March 21, 2014

Lt.Col Spaid is taking over at the CO of VMM-365, the Blue Knights, later this Spring. He started with Ospreys in 2005 and his background is with CH-46s. He was deployed on the first MEU, which worked with the Osprey in 2009, and now he is returning to sea with the 24th MEU this Fall.

In the interview, he discussed his experience with how the USMC introduced the Osprey, and has evolved the capability over time, and how the Osprey experience is being leveraged to prepare for the introduction of the CH-53K.

The pattern of innovation seen with the Osprey should almost certainly be repeated with both the F-35B and the CH-53K, whereby the assets are introduced, baby steps taken to get use to the aircraft, then to evolve the tactical understanding of the asset, and adjust the software to the evolution of tactics as well.

And in the course of all this, the fly by wire airplanes and their digital systems will allow a new approach to maintenance and mission planning.

LtCol Spaid during the 2014 interview.

LtCol Spade described his first deployment with the MEU as follows: *We gently eased in the aircraft, and learned its capabilities and built confidence in the aircraft and began to shape new tactics for the use of the aircraft.*

As he was preparing for the engagement with the 24th MEU, much has changed.

We now have matured tactics and the aircraft itself. We are leveraging the range, speed and capabilities of the aircraft as a ground assault vehicle.

During all of this evolution, the software has changed as well. He described the Osprey as a highly computerized and automated aircraft. He described the major shift from the CH-46 as a mechanical system to the Osprey as a digital one.

Spaid underscored: *A key element of the transition from the CH-46 to the Osprey as one of a significant shift in cockpit management and systems with the digitalized Osprey.*

Software in the aircraft has evolved over time and if you have been out of the cockpit for 24 months, you come back to the squadron for retraining.

The software evolves and with each upgrade, new publications are released explaining the changes and preparing the pilots and maintainers for the next phase.

By and large, the upgrade process is relatively seamless, but obviously some upgrades are more drastic than others.

The aircraft has improved over time as the automated systems perform better and the operators gain more confidence in the performance of the aircraft.

Spaid emphasized: *The ice protection systems are now better and a major change is the coupled mode whereby we fly close to the ground can really leverage automation. We can fly from very far out and hover off the ground by using the automated flight systems. This is a major operational advantage when coming into the objective area in supporting the Ground Combat Element.*

A key aspect of the impact of automation and digital systems is how maintenance is conducted.

According to LtCol Spaid: *The maintenance profile has shifted from mechanics to avionics. And the plane now has a flight history which is downloaded and the maintainers can directly examine the flight performance to determine repairs which need to be done.*

Another aspect of having digital systems tracking the flight envelope and flight experience is the ability to turn that data over to mission planners who can then use that information to plug into mapping software to determine the nature of the operational trajectory and battlespace experience surrounding the actual flight experience of that aircraft.

THE 2021 INTERVIEW

When the Osprey was introduced into combat in 2007, Col Spaid was part of that first squadron and indeed with his co-pilot were the first to enter Iraq flying from the Middle East.

His CH-46 squadron transitioned to Ospreys and he flew with that squadron to Iraq. He had flown with the HMM-263 in Iraq and was aware of its limitations and vulnerabilities, and saw the Osprey as providing capabilities for a different operational approach in Iraq.

According to Col Spaid: *We could now operate at altitude with speed and range and able to circumnavigate the battlespace and then insert in a favorable point into that battlespace. We also had a more survivable aircraft with the new materials used for the air frame as well.*

The airplane was so different from legacy aircraft, that he jokingly compared his transition to the Osprey as equivalent for him of being inconceivably selected for the space program, with new technologies, new capabilities, and very different operational possibilities.

The tiltrotor capabilities were certainly and still are revolutionary, but the maneuverability side of the aircraft was also a challenging part of the Osprey revolution.

Spaid underscored: *The fly by wire capability and unique flight control system of the aircraft was new for the legacy rotary-wing community. And learning to fly it and get used to what it could do was exciting and a challenge to transition from the old ways of doing things. When I started working precision landings, it took a couple of days to adjust. It was not a normal aircraft and doing precision landings was different as well.*

Indeed, when I was first learning how to fly the aircraft in the simulator and even first flights in the aircraft, I believed that it was going to be an area weapon. There's no way I'd be able to land precisely where I wanted to, but soon learned to do so with ease.

Then there was the tactical adjustment.

Being able to operate high and fast while minimizing our time in the climb and dive then coming in unexpected, that was a tactical advantage that no one else had or seen.

As the Marines were learning to use the aircraft in combat, my own observation was that Inside the Beltway, the aircraft started to get support from top leaders because it could take them around all of Iraq for inspections in a day, rather than having to operate over longer time within the limitations of how far a helicopter could fly.

Col Spaid noted that indeed he had that experience of flying senior military around Iraq and doing so as I described it. For example, *On Christmas Day in 2007, we flew General Petraeus around to three or four different FOBs in a single day, and he loved the plane.*

Col Spaid described the operational difference from a CH-46 in these terms: *A year and half earlier, I would fly a CH-46 from al-Assad out to al-Qa'im once a week. For rotary wing assets, you are going out West and the weather could be bad, and it was just a long-legged journey. With the Osprey, we did it daily.*

The Osprey started small in terms of deployment numbers for sure, but there was an esprit de corps to the team that would lead them to call themselves the Osprey Nation. It started by bringing back lessons learned from operations to then shape the TTPs for the follow-on squadrons.

As Col Spaid put it: *In Iraq we spent a lot of time in the dust doing reduced visibility landings. We worked some initial tactics and procedures and brought those back to the squadrons. And that started an earnest process of refining tactics.*

We were a small community, but we came from a variety of aviation settings and platforms. And that mix of different experiences informed our approach on how best to operate the Osprey. We had a very good mix of healthy work ethics which drove innovative thinking.

It was a melting pot of Marine Corps aviation. We all brought our best professional military aviator qualities into this effort, which means we had a unique opportunity to filter out bad habits that may have been lingering in our previous communities. We were working objective area mechanics and tactics for the aircraft and learning to fly the aircraft in those settings.

But as the community grew, standardization needed to be shaped.

In the period from 2010-2012, we focused heavily on standardization as West Coast squadrons were standing up.

U.S. Marine Corps Lt. Col. John W. Spaid VI, left, commanding officer, Marine Medium Tiltrotor Squadron 365, speaks with a Fox News reporter in regards to the MV-22 Osprey on Marine Corps Air Station (MCAS) New River, N.C., April 11, 2016. Reporters from Fox News visited MCAS New River. 2nd MAW, April 11, 2016.

Then in 2009, he deployed with the first MEU for the Osprey. Col Paul "Pup" Ryan was their squadron commander. They operated in the CENTCOM area with Fifth Fleet. They were in Bahrain, Kuwait and Iraq during the first MEU tour for the Osprey. This was the first time the aircraft operated from the sea base to project power from the sea.

Col Spaid served as the Aviation Maintenance Officer for that deployment. Among the 29 aircraft assigned to the composite squadron, 12 Ospreys were deployed with eight on the LHD and four on the LPD.

This was a learning challenge for the Navy as they had to adjust to Osprey operations and learn how it could operate from the ship, just as the Marines were learning how to maintain the new aircraft afloat. Eventually, the Navy officers onboard learned, in Spaid's words, *that we could fix it faster, we could launch it faster, we could fold it faster than originally expected. In the end, the confidence of the Navy officers and crew grew.*

This experience clearly impacted CENTCOM leadership for it set in motion what would become known as Special Purpose MAGTFs, and in EUCOM and AFRICOM is now called the North Africa Response Force, the NARF. Combatant Commanders learned, in Spaid's words: *that we could offload in Kuwait but operate all throughout Iraq. That was an eye opener for them.*

In effect, with the Osprey, the Marines were finally demonstrating what a long range shipboard assault support asset could do while the entire concept of an ARG-MEU was transforming.

Then in 2010, the squadron took the Osprey to Afghanistan, where LtCol "Buddy" Bianca was the commander, and I interviewed him prior to Afghanistan and after the first operations of the Osprey in Afghanistan as well.

This initial experience clearly has made Col Spaid a plank holder in Osprey nation.

As Spaid put it: *Being a plank owner and setting up a squadron is one thing but taking that aircraft with the first squadron and doing the first MEU, doing the first combat deployment in Iraq, you're not just the plank owner, you're driving the ship that you built the plank on. It is liking being a plank owner for a revolutionary military advantage.*

But he noted that at the time you really did not focus on that. *We were just doing our jobs. You focus on mission accomplishment; you don't really understand the historical significance of the event at the time. There were some really significant contributions there from the whole team, but I don't think you really appreciate it until later.*

When I first visited the flight line at New River in 2007, there were four aircraft on the tarmac. When looking at the tarmac, now there are a large number of Ospreys at rest, at least for a brief period of time between flights and operations. 13 years has brought significant change both to the Marine Corps and to the airplane and now will do so for the Navy as they introduce their variant of the Osprey.

As Col Spaid put it: *All our Marines are smarter. You still have hard working maintainers fixing the planes, but they are better armed with experience. Baseline pilot intellect is now through the roof. We've normalized what we thought was creative thinking and training. And now we're asking them to be even more creative and train for the next strategic challenge.*

I asked him, what was a primary focus during his time as CO of MAG-26 with regard to the aircraft itself?

For Col Spaid the answer to that question is sustainment, sustainment, sustainment. *A good number of the software upgrades we are making are with regard to reliability and benefit the maintainer and the maintenance process. We are working hand in hand with the program office and with industry to find the best way to keep this aircraft reliable.*

We then discussed the decision by the Navy to buy the Osprey for its resupply role.

Col Spaid highlighted several advantages.

- First, there was enhanced prioritization to sustainment of the aircraft throughout the entire Naval Aviation Enterprise.
- Second, was the benefit of training, as the Navy pilots are trained at New River and as the West Coast squadrons are stood up, there will be refresher training opportunities for Marines on the West Coast as well.
- Third, with the Osprey deployed to both the carrier and amphibious fleets, there will be greater opportunity for parts availability across the fleet.
- Fourth, there will be shared opportunities for upgrading the aircraft as well.

The initial training for USAF Osprey pilots and maintainers is at New River as well, but because the specialized mission and equipment on the USAF variant, there is less impact from commonality than will occur with regard to the Marines and the U.S. Navy working together, notably with the new emphasis on ways to shape enhanced Navy-Marine Corps integration as well.

But one area of partner training has occurred as the Japanese have acquired the Osprey and their pilots and maintainers trained at New River. For Col Spaid, the Japanese maturation at New River has been very significant and prepares the future for new FMS partners.

Spaid highlighted the opportunity: *As new FMS opportunities arise, we'll be able to leverage off the great success we have had with our Japanese allies. We'll change some things up a little bit, but it was a really good experience.*

In short, Col Spaid was on the ground floor with the birth of Osprey nation. And he is driving change as the next phases of Osprey operations unfold.

And his story is a personal embodiment of the growth of tiltrotor experience, which is a focus of my companion volume to this book to be published in the Fall of 2025.

3

A TILTROTOR ENTERPRISE

The concept of having a combat "helicopter" that can fly at the speed of an airplane has been around for a long time. The challenge was actually to design and build one that worked and was robust enough to operate and be sustained in combat.

And that was a major challenge. When the MV-22B was first introduced into Iraq in 2007 and a few years later the Air Force Special Forces introduced their variant of the Osprey, the CV-22, the U.S. military started to experience what such a capability could provide the operational force.

But to do so required the emergence of a team of operators, maintainers, industrial support and government engagement that made such an effort feasible.

And as the forces gained operational experience in a diversity of settings, a tiltrotor enterprise emerged. And as the U.S. Navy developed its variant of the Osprey and as the Army committed to a new tiltrotor aircraft for its forces, the enterprise moved into the new era of multi-domain operations in a world of peer competitors.

But this did not appear overnight, and it has been anchored

by a core team which has provided for the ongoing evolution of con-ops and missions and capabilities.

As Dan Gouré put it in a 2021 article: *It has been three decades since the V-22 Osprey first flew. Over that time, the V-22 has confounded its critics and more than proven itself in operations from Southwest Asia to the Western Pacific. Its abilities to fly like an airplane and land like a helicopter are particularly well- suited to an era of distributed operations...*

One reason the V-22 has succeeded is that it was given the time to work out technical problems. It also had a dedicated champion, the U.S. Marine Corps, that saw the value in the Osprey and worked hard to figure out how to operate a new kind of flying machine.[1]

The team quality of the effort has been highlighted in a NAVAIR article published in 2022 on the occasion of the 40 years of collaboration in developing, evolving and supporting this capability:

Forty years ago this month, the Department of Navy (DoN) took control of what is now known as the V-22 Joint Program Office (PMA-275), responsible for the cradle-to-grave acquisition, sustainment, development and production of the venerable tiltrotor aircraft.

With more than 700,000 flight hours, the V-22 Osprey is a military marvel, providing unmatched capabilities to the U.S. Marines, Navy, Air Force and the Japan Ground Self-Defense Force.

"Not a single flight hour, from the first to most recent, would have been possible without the leadership, innovation and partnerships developed in this joint program office," said Col. Brian Taylor, PMA-275 program manager.

"As the thirteenth leader in this role, I walked through the door to a well-established and exceptional team, cross collaborating to ensure the V-22 remains ready, reliable, relevant and safe through the 2050s."

Commemorative poster created for the 40th anniversary of the
Navy's V-22 program office, established originally as the
advanced vertical lift (JVX) program and today, known as the
V-22 Joint Program Office (PMA-275). Designed by Alyse
Joseph, PMA-275 Communications Specialist and Graphic
Designer.

Following the failure of Operation Eagle Claw *in 1980 – the
attempted and then aborted mission to rescue 53 U.S. embassy staff
members in Tehran – the Defense Department saw the need for an
aircraft that could support long-range, high-speed missions utilizing
vertical take-offs and landings. As a result, the department initiated the
establishment of the advanced vertical lift program.*

In December 1982, executive leaders transferred the newly formed program, originally led by the Army, to the DoN and established the Joint Services Advanced Vertical Lift (JVX) program. A few years later, it would become the V-22 Osprey program.

As a first of its kind, the V-22 came with a complex development and testing program, integrating unprecedented technology and propulsion elements.

Following first flight in 1989, the development program continued to refine the design with the Marine Corps variant, MV-22, beginning operation in 2000 and fielding in 2007. Not long after, in 2009, the Air Force declared Initial Operating Capability for its variant, the CV-22.

Over the last 10 years, the V-22's aperture widened, welcoming the U.S. Navy (CMV-22B) and Japan Ground Self-Defense Force to its portfolio, increasing the aircraft's global impact.

Today, the Osprey's mission has grown and includes medium-lift assault support, long-range infiltration / exfiltration, at-sea cargo resupply, combat logistics, medical evacuation and more.

By acknowledging the operational success of the V-22, the program also recognizes the challenges and adversity faced throughout its development, all providing the lessons and experience required to build and maintain the aircraft's relevance for decades to come.

"The warfighters who fly, maintain and rely on the aircraft, deserve nothing less," said Taylor. "As a program, we keep those lost during mishaps in our memory; their sacrifice to this nation cannot be overstated."

As a joint program office, PMA-275 works with representatives from all service branches and its international partner, Japan, that fly and maintain the aircraft.

In addition, it works closely with its industry partners, from original equipment manufacturers Bell-Boeing, Rolls-Royce, and Raytheon to the hundreds of suppliers keeping the aircraft flying.

Keeping these partnerships strong, both within the Defense Department and industry, ensures that all V-22 stakeholders are informed and ready to work together on all aspects of the program, from emergent to day-to-day tasks.[2]

One way to conceptualize the timeline for the emergence

and evolution of the tiltrotor enterprise is by the platforms which have been developed and operated by the various services since the introduction of tiltrotor capability.

By declared initial operating capabilities (IOCs), the four platforms associated with this enterprise go from the MV-22B to the CV-22B to the CMV-22B to the FLRAA or Future Long-Range Assault Aircraft.

MV-22B: IOC 2007

This is the backbone of the enterprise. The pioneer platform and with the USMC driving its development and concepts of operations, it provided the breakthrough for the use of this innovative technological capability.

A U.S. Marine Corps MV-22 Osprey tiltrotor aircraft with Marine Medium Tiltrotor Squadron (VMM) 262, Marine Aircraft Group 36, 1st Marine Aircraft Wing, conducts a bilateral formation flight alongside Japan Ground Self-Defense Force service members with Western Army Aviation Group during the field training exercise portion of Resolute Dragon 23 over Kumamoto, Japan, Oct. 18, 2023. U.S. Marine Corps photo by Cpl. Kyle Chan.

This is how NAVAIR describes the MV-22B: *The MV-22B Osprey is a tiltrotor V/STOL aircraft designed as the medium-lift replacement for the CH-46E Sea Knight assault support helicopter. The Osprey can operate as a helicopter or a turboprop aircraft and offers twice the speed, six times the range, and three times the payload of the CH-46E.*

Initial Operational Capability (IOC) for the MV-22B was declared in June 2007. The Osprey had three successful combat deployments in Iraq from October 2007 to April 2009 with VMM-263, VMM-162 and VMM-266 respectively. VMM-263 embarked on the first MV-22 shipboard deployment with the Bataan Ready Group in May 2009 as part of the 22nd Marine Expeditionary Unit (MEU).[3]

THE CV-22B: IOC 2009

The CV-22B is the specialized variant of the Osprey designed to support Air Force special forces operations.

This aircraft has always interested me in part because of my relationship with Dr. Zbigniew Brzezinski. He was my main professor at Columbia University when working on my PhD and then my boss at the Research Institute of International Change at Columbia. When he went to Washington, I would follow and we stayed in contact afterwords.

Of course, he was one of the key policy makers launching the unsuccessful attempt to rescue U.S. hostages in Tehran. Ironically, the Marines would develop a capability to do this type of mission with the SP-MAGTF.

I remember a meeting with Brzezinski where I described the SP-MAGTF and its function (which was shaped after the Benghazi situation) and which he felt if such capability was available when he was NSC advisor, the mission would have succeeded.

This is how NAVAIR describes the CV-22B: *The CV-22 is the Air Force Special Operations Command (AFSOC) variant of the U.S. Marine Corps MV-22 Osprey. The CV-22's mission is to conduct long-range infiltration, exfiltration and resupply missions for special operations forces.*

Like the MV-22 Osprey, the CV-22 is a tiltrotor aircraft that combines the vertical takeoff, hover, and vertical landing qualities of a helicopter with the long-range, fuel efficiency and speed characteristics of a turboprop aircraft. Those capabilities give this versatile, self-deployable aircraft the capability to conduct missions that would normally require both fixed-wing and rotary-wing aircraft.

A 71st Special Operations Squadron CV-22 Osprey receives fuel from a 522 SOS, MC-130J Combat Shadow II aircraft, over the skies of New Mexico, Jan. 4, 2012. The 71 SOS is stationed at Kirtland Air Force Base, N.M., and conducted air refueling training with the 522 SOS, stationed at Cannon Air Force Base, N.M.

The CV-22 is equipped with integrated threat countermeasures, terrain-following radar, forward-looking infrared sensor, and other advanced avionics systems that allow it to operate at low altitude in adverse weather conditions and medium- to high-threat environments.

The first two test aircraft were delivered to Edwards Air Force Base, Calif., in September 2000. The 58th Special Operations Wing at Kirtland AFB, NM began CV-22 aircrew training with the first two production aircraft in August 2006. The first operational CV-22 was delivered to Air Force Special Operations Command in January 2007. Initial operational capability was achieved in 2009. [4]

THE CMV-22B: IOC 2021

The CMV-22b was selected by the U.S. Navy to replace the C-2A Greyhound for carrier support operations. But the versatility of the CMV-22B compared to the C-2A has opened a whole new way of providing for fleet support.

This is how NAVAIR describes the CMV-22B: *The CMV-22B is the Navy's long-range/medium-lift element of the intra-theater aerial logistics capability; it fulfills the Joint Force Maritime Component*

Commander (JFMCC) time-critical logistics air connector requirements by transporting personnel, mail and priority cargo from advance bases to the Seabase.

The CMV-22B Osprey is a variant of the MV-22B and is the replacement for the C-2A Greyhound for the Carrier Onboard Delivery (COD) mission. The Osprey is a tiltrotor V/STOL aircraft that can takeoff and land as a helicopter but transit as a turboprop aircraft.

PACIFIC OCEAN (Sept. 12, 2023) A CMV-22B Osprey from the "Sunhawks" of Fleet Logistics Multi-Mission Squadron (VRM) 50 and a CMV-22B Osprey from the "Titans" of Fleet Logistics Multi-Mission Squadron (VRM) 30 conduct flight operations on the flight deck of the aircraft carrier USS Nimitz (CVN 68). Nimitz is underway conducting routine operations. (U.S. Navy photo by Mass Communication Specialist 3rd Class Emma Burgess)

It will provide the Navy with significant increases in capability and operational flexibility over the C-2A. CMV-22B operations can be either shore-based, "expeditionary", or sea-based. The Osprey is a critical warfighting enabler, providing the time sensitive combat logistics needed to support combat operations.

As compared to the MV-22B, the Navy variant has extended operational range, a beyond line-of-sight HF radio, improved fuel dump capability, a public address system for passengers, and an improved lighting system for cargo loading. The CMV-22B will be capable of transporting up to 6,000 pounds of cargo/personnel to a 1,150 NM range.

43

The CMV-22B declared Initial Operational Capability (IOC) in 2021.[5]

THE FLRAA OR FUTURE LONG-RANGE ASSAULT AIRCRAFT: DOWNSELECT BY THE U.S. ARMY IN 2022

On December 5, 2022, the U.S. Army down selected Bell's new tiltrotor aircraft to provide for its new vertical lift capability.

The U.S. Army has labeled this as the FLRAA or Future Long-Range Assault Aircraft.

Army Chief of Staff Gen. James McConville in 2023 noted that the Bell aircraft cruise speed of more than 280 kt. and combat range of more than 500nm will be important for future operations in the Indo-Pacific.

What they've also done well is to get the aircraft more maneuverable at the destination. It's good that you can go long range. It's good that you can go fast. But then you have to be able to do the type of operations that we do — maneuver at the destination.[6]

In a December 2023 article published by the U.S. Army, the FLRAA was described this way:

A tiltrotor aircraft, FLRAA will have the hybrid capabilities of planes and helicopters. When fielded, it will expand the depth of the battlefield by extending the reach of air assault missions and enable ground forces to converge through decentralized operations at extended distances.

FLRAA's inherent reach and standoff capabilities will ensure mission success through tactical maneuver at operational and strategic distances while the aircraft's speed and range will nearly double the Army's patient evacuation capability during the "Golden Hour."

The FLRAA is intended to eventually replace part of the U.S. Army's UH-60 Black Hawk fleet, which has been in service for more than four decades.[7]

The Army is inheriting the experience of the other services in their use and the evolution of the tiltrotor enterprise. Tiltrotors provide warfighting commanders and diplomatic missions across the globe operational lethality, flexibility, and surprise through the speed, range, payload, and survivability only a tiltrotor aircraft can provide.

Tiltrotors allow commanders to respond rapidly to crises at lower risk to friendly forces and lower diplomatic burdens and can do so across great distances.

And in describing the purpose of the FLRAA, a U.S. Department of Defense publication underscored: *Units will utilize FLRAA's increased speed, range, and maneuverability to assault enemy forces from areas of relative safety outside the range of enemy long-range fires. The FLRAA will be effective, decisive, and survivable in the lower tier of the air domain. It will integrate other programs within the FVL ecosystem.*[8]

This is a new-build tiltrotor incorporating new technologies and capabilities designed to keep affordability and sustainability costs down and should prove attractive for allies as well. In fact, the U.S. Army has a very proactive approach to involving allies in the program.

In an October 2022 article by Meredith Roaten, the Army's focus on allied engagement in the program was underlined: *The Army's team for next-generation helicopters and drones wants to bring more allies into the development process for its major programs, an official said Oct. 10.*

The United Kingdom and the Netherlands already have cooperation agreements with the Army, said Maj. Gen. Walter Rugen, director of the future vertical lift cross functional team. But the service is working to secure additional ways to bring its allies into the fold....

Australia's military has an embedded exchange officer in the future vertical lift cross functional team, which has led to "the Aussies working hand in glove with us," he added.[9]

Looking back and looking forward, one might note that the V-22 is a large amphibious connector often used for assaults and everything else. The FLRAA is a purpose-built assault aircraft with the requisite reliability and affordability improvements to make a large fleet sustainable for the Army.

Even if the Army stops at its initial 573-aircraft buy, which it is unlikely to, their fleet will rapidly eclipse the size of the Marine, Navy, Air Force, and Japanese fleets combined. It is a new phase indeed for the tiltrotor enterprise, and not just in terms of delivering a new variant.

4

THE TIMELINE FOR THE TILTROTOR ENTERPRISE: A CON-OPS PERSPECTIVE

A nother way to look at the timeline for the development of the tiltrotor enterprise is from the perspective of the evolving concepts of operations and mission profiles which have been developed by the operators of the aircraft.

And as each phase of the learning curve was mastered that provided a foundation for another phase of the learning curve. The tiltrotor enterprise did not experience the classic challenge of mission creep: it was an enterprise that embraced the expansion of missions as new operational lessons were learned and new capabilities could be added.

A timeline conceptualized for MV-22B operations in these terms might look like this:

THE MV-22B: PHASES OF OPS DEVELOPMENT

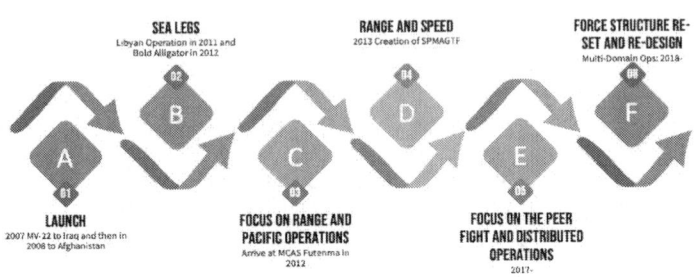

I will provide a brief characterization of each phase in the con-ops evolution time line and will then provide more detailed looks at each phase of development in later chapters.

LAUNCH: 2007-2011

The Osprey was IOCd and then sent to Iraq. And then in 2009 it was sent to Afghanistan.

The Marines who took the Osprey were pioneers and original stakeholders in what would become a tiltrotor enterprise. But they were taking a new aircraft into combat: it was a test by fire, not flights back in test ranges in CONUS.

And operating in Iraq and Afghanistan in terms of the terrain was very different. The aircraft was being challenged to perform in very different terrain and climatic conditions. And the Marines were part of a joint force operating against a reactive enemy.

The speed and range capabilities of the Osprey became evident early on in Iraq. As one Army general I interviewed who flew on the Osprey to visit his troops scattered throughout Iraq put it to me with regard to why he chose the Osprey on which to fly:

It was the only aircraft that could cover Iraq and land in the various locations I had to visit in one day.

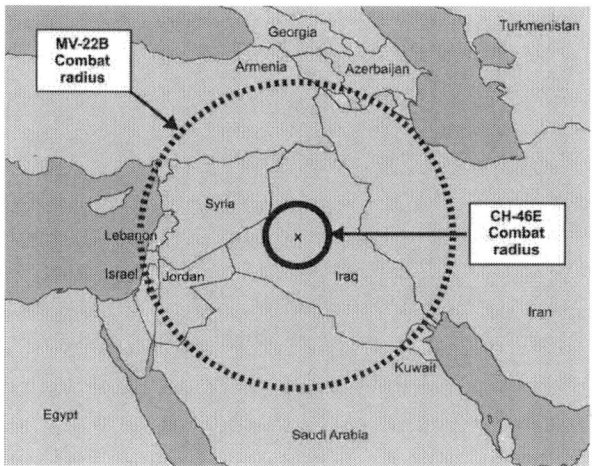

USMC graphic showing MV-22B coverage capability in Iraq.
2008

The ability to fly and land at speeds and angles which a rotorcraft could not do allowed the other elements of the USMC force to be used differently from the outset in Iraq. And when it went to Afghanistan, it would enable the development of new concepts of operations in using the rotorcraft as well.

As I noted in 2012 overview piece on this period: *We can start first with the decision of USMC leaders to deploy the plane to Iraq. This deployment was itself part of the "testing" process. What is often overlooked is that testing is really done by pilots and maintainers in combat, not by technicians in white coats or statisticians at the GAO.*

There was clear concern expressed to me by Marine Corps Aviators that the deployment to Iraq would prove challenging, and its was. But it was also evidence of the role of leadership in making the hard decisions to role out needed capabilities and let the users define the direction of a program, not the program managers....

What we have seen is that the plane started with "training wheels

49

on" its deployments and those wheels not only have been thrown off, but as time in combat has gone up, the Marines as well as the Combatant Commanders have begun to understand what a transformational platform can do when connected with other capabilities and assets.

A V-22 Osprey lands at Baghdad International Airport Saturday, Oct. 27, 2007. A fleet of 10 Ospreys arrived in Iraq earlier in the month. The tilt rotor Ospreys have the capability to take off and land like a helicopter and fly like a fixed-wing aircraft. Marines are using the Osprey primarily in Anbar province. October 27, 2007. Photo by Staff Sgt. Curt Cashour.

The plane started in Iraq built around a famous diagram showing the speed and range of the aircraft in covering Iraq. As one Marine commented: "The MV-22 in the Area of Operation (AO) was like turning the size of the state of Rhode Island into the size of Texas."

It was the only "helicopter" that could completely cover Iraqi territory. And in this role, the testing of support as well as operational capabilities was somewhat limited as Marines tested out capabilities and dealt with operational challenges. The plane was largely used for passenger and cargo transport in support operations in difficult terrain and operating conditions.

It was used for assault operations from the beginning but over time, the role would expand as the support structure matured, readiness rates grew and airplane availability become increasingly robust.

From the beginning the aircraft impressed and foreshadowed later developments.

As General Walsh, *now Deputy Commanding General Marine Corps Combat Development Command noted in my interview with him at the Cherry Point Air Station in 2009 after a year in Iraq that with the withdrawal of U.S. forces from Iraq there was a roll up of forward operating bases.*

This meant that the remaining forces had to cover more ground and to provide protection at greater distance. Enter the Osprey, which did not require Forward Operating Bases (FOBs) to provide lift and support to forward deployed forces.[1]

Iraq was the beginning and a consciousness raiser for troops and commanders.

Next on the agenda was the beginning of deployments to Afghanistan, which of course continue.

The Afghan phase of deployments has seen the aircraft and its operator's transition to more assault combat operations over time, to the point where the latest Osprey squadron just came back from Afghanistan with record setting assault operations for the Osprey.

Osprey operating in Afghanistan in 2009. Credit: LtCol Biancca

In the words of the head of 2nd Marine Air Wing – Major General Glen Walters — upon his return from Afghanistan:

"The Ospreys had their normal fair share of general support, resup-

plies, etc. But we started accelerating their use as my time there went on, and used them for both the conventional and Special Forces operations.

"The beauty of the speed of the Osprey is that you can get the Special Operations forces where they need to be and to augment what the conventional forces were doing and thereby take pressure off of the conventional forces.

"And with the SAME assets, you could make multiple trips or make multiple hits, which allowed us to shape what the Taliban was trying to do.

"The Taliban has a very rudimentary but effective early warning system for counter-air. They spaced guys around their area of interest, their headquarters, etc.

"Then they would call in on cell or satellite phones to chat or track. It was very easy for them to track. They had names for our aircraft, like the CH-53s, which they called "Fat Cows."

"But they did not talk much about the Osprey because they were so quick and lethal.

"And because of its speed and range, you did not have to come on the axis that would expect. You could go around, or behind them and then zip in. We also started expanding our night operations with the Osprey. We rigged up a V-22 for battlefield illumination.

"A lot of these mission sets were never designed into the V-22 but you put it into the field and configure it to do the various missions required. And we have new software for the Ospreys in Afghanistan where you can pick your approach, angle, approach speed and let the aircraft do it all. That is a huge safety gain."[2]

SEA LEGS: 2011-2015

Without success in the first phase, there would not have been the next one. The Marines operate from flexible and mobile bases. Sea basing is a key attribute of the Marines as the nation's 9/11 force. But the operation of the Osprey from the sea was not the primary use of the aircraft when introduced with the priority being placed on the land wars.

But the Marines early on focused on developing their sea legs and learning how to sustain the aircraft afloat. These are different skill sets than operating from land and land base support elements.

I attended several of the Bold Alligator exercises in this period and saw and flew on Ospreys which were a key part of developing a new approach to assault at distance from the sea. I remember Bold Alligator 2012 where the journalists were waiting for the amphibious vehicle assault from the ship but the assault had already happened deep inland by an Osprey contingent.

And during the exercise an Osprey landed on an Military Sealift Command TAKE ship, demonstrating its flexibility to work within the fleet.

An MV-22 Osprey assigned to the Fighting Griffins of Marine Medium Tiltrotor Squadron (VMM) 266 makes a historic first landing aboard the Military Sealift Command dry cargo and ammunition ship USNS Robert E. Peary (T-AKE 5). The Osprey landed aboard Robert E. Peary while conducting an experimental resupply of Marines during exercise Bold Alligator 2012. Credit: USMC

But the hallmark event in this period was the Libyan intervention. Here the Obama Administration decided to "lead from behind" which included not using a large deck aircraft carrier in the operation but sending an ARG-MEU. But this ARG-MEU was Osprey-enabled and did things no ARG-MEU had ever done before.

This became clearly evident when a USAF pilot had to bail out in enemy territory and the USAF reaction time paled in comparison to what the Marines could do with the Osprey and the Marines pulled the pilot out before he could be captured and might have become a political pawn in the conflict.

As my long-time colleague Ed Timperlake wrote in 2011: *A key military and diplomatic result can occur from an action, which eliminates a bad result. In the social media or internet age, a problem that can be seen is more likely to be discussed than a problem which did not happen because of a military success.*

More often than not, core military professionalism enhanced by the appropriate equipment actually works and problems are avoided by NOT having a failure.

In a certain sense, the phenomenon of avoiding an outcome by having a capability to create a better outcome is central to military or diplomatic success.

It is important to indicate the value of a platform, system or professional capability, which shapes avoidance of a negative outcome, as well as describing in terms of what it does. What a system can avoid is in many ways as important as describing what it can do.

In some ways, this is the military variant of a null hypothesis. One can prove the validity of the military platform, system and capability by what it avoids. One tests its value by hypothesizing alternative outcomes absent the capability.

I once had an enjoyable discussion with Ron Maxwell the Director of the great American epic movie Gettysburg. His art and vision will truly stand the test of time. I was curious about all of his many challenges. He said one of the most difficult for both a Director and the actors in such a movie was to stay in the moment as it was historically playing out.

In other words, it was important not to act as the outcome was already pre-determined. The actors playing real life battlefield commanders had to assume they were making life and death decisions for thou-

sands of lives and the fate of America without certainty of the outcome. That was the great success of the movie since both the Director and actors truly captured the moment-to-moment uncertainty of decision making in an epic historical battle.

In real life, commanders are constantly faced with making decisions, some routine and in fact boring others as Gettysburg shows are decisions that can have huge consequences. It is why experienced leaders, not managers, are needed at all times in the American military and is one of the core differences between real commanders and cubical commandos.

A leader knows what to do in the moment of action. Getting to that moment requires preparation and attention to detail while inspiring all to perform their crucial responsibilities to the best of their ability. Training tactics and technology have to all come together to accomplish the mission.

The pilot and flight crew of an MV-22B Osprey with Marine Medium Tiltrotor Squadron (VMM) 266 Reinforced which was involved in the rescue of a downed Air Force pilot during Operation Odyssey Dawn received Air Medals with the combat distinguishing device for valor at Marine Corps Air Station New River, N.C., Jan. 7, 2013. VMM-266 is again reinforcing the 26th Marine Expeditionary Unit for their 2013 deployment. U.S. Marine Corps photo by Cpl. Michael S. Lockett.

Of course, a reactive enemy and the "fog of battle" always comes into play.

Finally, Napoleon's quip about wanting a "lucky" General is more real then people think and good commanders can both make their luck and also catch a break with a reactive enemy making a mistake. But they do not have perfect knowledge.

When a USAF F-15E Strike Eagle crew call sign "Bolar 34" ejected over Libya in the current NATO air war against Gaddafi, they were in real difficulty without a trained rescue team. And a U.S. Navy Marine team off the Libyan coast was trained and ready to save them.

The USS Kearsarge Amphibious Readiness Group with Navy Commodore Pete Pagano Amphibious Squadron Four, and Col. Mark Desens's 26th Marine Expeditionary Unit were in the right place at the right time.

The time of the operation was significantly reduced because of the team and their use of the Osprey aircrafts onboard. The time from order to execute to the 26th MEU's MV-22 successful extraction, called a TRAP (Tactical Recovery of Aircraft and Personnel) recovery mission was 47 minutes. Those forty-seven minutes may have changed the entire narrative of ongoing combat operations against Qaddafi.

From Gettysburg and other bloody Civil War Battles, America has shown as a nation we will accept horrendous causalities if the individual warrior and their families think the cause is just. However, there is another American core value as exemplified by the American military saying, "leave no one behind." The TRAP mission captured that battle-field ethos.

However, what if the aircrew hadn't been saved but rather fallen into the hands of Gaddafi's forces? Current events would be way different. The leader of Libya maybe a lot of horrendous things but over time he knows how to play the propaganda game as a master. One American in enemy hands, if history is any guide, would have pivoted the world wide 24/7 media to a "POW Story."

Whatever one thinks of the political decision to have a NATO Libyan Air War, it is being fought by mature adults who saved the day in the TARP mission. An Administration can be vexed and paralyzed by a POW story line. President Obama can be thankful of not having to face the null hypothesis which would have resulted from NOT having the TRAP team armed with the Osprey....

Now with the horrific weapon and the will to use it by Islamic fanatics of filming captives being beheaded, the issue of being held captive has escalated to an entirely new level of media attention. One U.S. or Allied airman captured will certainly get attention and perhaps even change the course of events.

The USN/USMC ARG-MEU team did not let the Obama Administration have that dilemma, at least for now.

So how did the Marine Commander accomplish his mission?

He built on the dedication and vision of those who came before. Land based assets in support of the Libya operation were time, distance and tanker constrained. But the Marines had both the MV-22 and the VSTOL AV-8 Harriers on the ARG deck.

In ten minutes from the sea, Harriers could go from feet wet to feet dry. The Osprey could swoop into any terrain hover land and extract the crew. CH-53's with a strong quick reaction combat force of Marine infantry could be called into battle at a moment's notice. This aviation technology revolution was the key along with Marine training going back to the founding of the Corps.

The ARG was afloat in the Med while the normal USN Med Carrier Battle Group this one organized around the USS Enterprise "The Big-E" was in the Indian Ocean flying air-to-ground combat missions in support of NATO forces fighting in Afghanistan.

The "Big-E" is a Navy legend for over four decades. In peace and war ever ready for what ever the National Command Authority thinks necessary.

But if one were to watch the Enterprise in action from Yankee Station off the Coast of Vietnam in the sixties to the current ops in the IO in support of Afghanistan mission there would be little difference except more modern fighter and attack aircraft. In fact the F/A-18 brings both missions together.

However, if one were to observe the Navy/Marine Gator team off the Coast of Vietnam in 1968 to Col. Dessen's 26th MEU it would not be recognized.

The emerging ARG with the ACE (Air Combat Element) of the MV-22 and VSTOL Harrier – soon to be followed by the revolutionary Air-to-Air and Air-to-Ground VSTOL F-35B will revolutionize American combat agility from the Sea to the Sea for decades to come. Of this there can be no question.

The Amphibious Ready Group is becoming the Agile Response Group under the influence of the new technologies being deployed on the ARG ships.

As Marines are want to say to Navy Carrier Battle Group's "good on em" but please pay attention to the fact that in one brief moment, 47 minutes, Marine air assets altered the course of our most current war.[3]

FOCUS ON RANGE AND PACIFIC OPERATIONS: 2012-

The MV-22B and the USMC would face a unique situation when the Obama Administrating announced their "shift to the Pacific" in 2011.

According to a 2012 Congressional Research Service Report:

In the fall of 2011, the Obama Administration issued a series of announcements indicating that the United States would be expanding and intensifying its already significant role in the Asia-Pacific, particularly in the southern part of the region.

The fundamental goal underpinning the shift is to devote more effort to influencing the development of the Asia-Pacific's norms and rules, particularly as China emerges as an ever-more influential regional power.[4]

Despite this announcement, the priority was on the "good war" in Afghanistan, and the Administration did not put new resources and capabilities into the Pacific.

Rather, they had cancelled the F-22 program and put a hold on the F-35B program and had no upsurge in shipbuilding which would be needed for such a shift. The Osprey flew into the Pacific as the only new asset which presaged what would become what we face now, namely, a major force structure re-orientation on distributed operations in the Pacific.

Put another way, the Osprey flew into history as the new platform shaping what would become a major strategic shift in the current period. What the Osprey brought to the effort was a unique capability in terms of speed and range and landing flexibility to cover areas of interest for the U.S. military in terms of the insertion of force and of supplies.

The Ospreys first arrived in the Pacific in 2012. Given the situation in Japan this was a complicated arrival, given Japanese concerns with the aircraft. But through the years, confidence grew and the Japanese themselves would buy and operate the aircraft.

This article by Lance Cpl. Benjamin Pryer published on 23 July 2012 announced the Osprey's arrival: *Twelve MV-22 Osprey tiltrotor aircraft were off-loaded from a civilian cargo ship at Marine Corps Air Station Iwakuni, Japan, today. This marks the first deployment of the MV-22 to Japan. The aircraft will be stationed aboard Marine Corps Air Station Futenma in Okinawa, Japan, as part of Marine Medium Helicopter Squadron 265 (HMM-265).*

MCAS Iwakuni features both an airfield and a port facility, making it a safe and operationally feasible location to offload the aircraft. The offload was closely coordinated with Government of Japan.

"We are obviously pleased to demonstrate the capacity of this co-located deep water harbor and aerial port of operations. It clearly highlights Iwakuni's position as a logistical lynchpin in the strategic alliance between the United States and Japan here in the Western Pacific," said Col James C. Stewart, Commanding Officer of Marine Corps Air Station Iwakuni.

Marines will prepare the aircraft for flight after its 5000-mile journey aboard the civilian cargo ship Green Ridge. However, the MV-22 Ospreys will not conduct functional check flights until the results of safety investigations are presented to the Government of Japan and the safety of flight operations is confirmed. Following safety confirmation and functional check flights, the Ospreys will fly to their new home aboard MCAS Futenma.

Groups opposed to the MV-22 deployment in Japan have demon-

strated in Okinawa and Iwakuni. Recognizing the concerns of Japanese citizens led U.S. and Japanese officials to ensure safety of flight operations is confirmed before Ospreys fly in Japan.

Deployment of the MV-22 Osprey to Japan marks a significant step forward in modernization of Marine Corps aircraft here in support of the U.S. Japan Security Alliance. Throughout the Marine Corps, Ospreys have been replacing CH-46 Sea Knight helicopters, which made their Marine Corps debut during the Vietnam era.

The Osprey is a revolutionary and highly-capable aircraft with an excellent operational safety record. It combines the vertical capability of a helicopter with the speed and range of a fixed-wing aircraft.

The Osprey's capabilities will significantly strengthen the Marine Corps' ability to provide for the defense of Japan, perform humanitarian assistance and disaster relief operations and fulfill other Alliance roles.

The Osprey has assisted in humanitarian operations in Haiti, participated in the recovery of a downed U.S. pilot in Libya, supported combat operations in Iraq and Afghanistan, and has conducted multiple Marine Expeditionary Unit deployments.

As of April 11, 2012, the Osprey has flown more than 115,000 flight hours, with approximately one third of the total hours flown during the last two years.

A second squadron of 12 aircraft is scheduled to arrive at MCAS Futenma during the summer of 2013.[5]

The Marines were combining an operational shift which they called a "distributed laydown" which absolutely could not happen without the range and flexibility which the Osprey enabled.

Its long legs supported by aerial refueling was a new capability within a new strategic situation. Let me put this bluntly: With the CH-46, the platform the Osprey was replacing, this was not going to happen.

RANGE AND SPEED: THE CREATION OF SP-MAGTF, 2013

The shift to the Pacific was one strategic impulse which leveraged the evolving capabilities of the Osprey. The 2012 Benghazi attack was another.

Terrorists attacked the United States diplomatic compound and adjoining CIA Annex in Benghazi, Libya, on September 11, 2012. Despite repeated warnings from officials about the security risks in Tripoli and Benghazi, nothing had been done to prevent the attacks or to remove personnel.

Although we had seen something similar happen in Iran at the end of the Carter Administration, it was only the wake of the Benghazi attack that creating new capability built around the Osprey was a focus of attention. This would lead to the formation the SP-MAGTF.

In a 2013 article, I focused on the standup of the SP-MAGTF at 2nd Marine Air wing.

During this year's visit to New River, a discussion with Major Frank "Robo" Rhobotham, VMM-364 Remain Behind Element (RBE) Officer in Charge (OIC), underscored how significant the Osprey has been in forcing culture change in the USMC and shaping new combat approaches.

In the discussion with Rhobotham, two key aspects of change were discussed. The first involved the standpoint of the Special Purpose MAGTF, now in Spain, and the second the changing approaches associated with a younger generation of maintainers, who work on the Osprey.

Rhobotham discussed the very short period from the generation of the concept of the Special Purpose MAGTF to its execution. It took about eight months from inception to deployment.

He emphasized the flexibility of the force and its light footprint. "With a six-ship Osprey force supported by three C-130s we can move it as needed. The three C-130s are carrying all the support equipment to operate the force as well."

"The flexibility, which the Osprey now offers Combatant Comman-

ders and U.S. defense officials, is a major strategic and tactical tool for the kind of global reality the U.S. now faces, requiring rapid support and insertion of force.

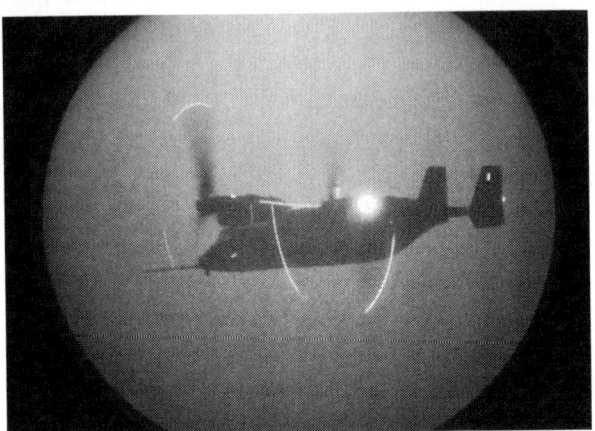

An MV-22B Osprey with Special Purpose Marine Air-Ground Task Force Crisis Response, prepares to conduct nighttime tiltrotor air-to-air refueling with a KC-130J Hercules over the southern coast of Spain, Aug. 22, 2013. U.S Marine Corps photo by Cpl. Michael Petersheim.

Question: Could you describe the process as seen from your end?

Major Rhobotham: "We received a request to look at the deployment of a 6-ship detachment of Ospreys to operate flexibly as a group. That started a long, long conversation because it depends on what you want to do with those six airplanes.

"Are we going austere, are we working from a prepared zone; are we going to fly 100 hours a month, or are we going to fly 500 hours a month? And where are we going to operate it for environment matters to the performance and endurance of the aircraft?

"It is similar to thinking about ground transportation and what car you would use., Take the Baja 500, for example. If you buy a truck from the Ford dealership, and you drive it around LA, you're going to get 150,000 miles out of it. You take that same truck and you attempt to run the Baja 500, you won't make it past the first day.

"It is the same thing with this aircraft. If I go from paved runway to paved runway and I fly in airplane mode the entire way, I'm going to get a lot more use out of it. And if I'm flying it like a helicopter and I'm landing in nasty, dusty, dirty environments, it's going to break down, as any mechanical device will do in that environment.

"Additional questions had to be answered. Where are we going? What are we going to do, how much are we going to fly? Who are we supporting, how are we being supported?

"We then got a response back that Africom was interested in what a six-plane V-22 force would look like. With the Africom focus that shrunk the bubble down. The continent of Africa has about every single environment out there.

"Mali happens in the middle of this. While we were not told that Mali was even in the play, it was dominating the news in the time period that we began the planning. We really started looking at the western coast of Northern Africa, we looked at the Northern portion of Africa, and obviously Libya is all fresh in everybody's memory.

"We're an assault support unit; we are always supporting, and we're supporting the Marines. So obviously, we, by ourselves, are not a force. We enable somebody else to be a more efficient, more effective force.

"And that also helps as well in thinking about the deployment focus.

"What can we do with the company, how can we help a company? And we fell back upon our MEU mission sets. If we're going to be supporting the African countries with a company, we draw upon what we know.

"We were given some restraints to the diplomatic clearances with our European partners, which shaped the force to a certain degree. And then there's always the what-ifs.

"As a result, we deployed out relatively heavy. We're running two ships of maintenance in the field, and we have round the clock maintenance."

Comment: Obviously the light footprint of the force gives it significant operational flexibility.

Major Rhobotham: "That is a significant benefit. If for some reason, *due to political turmoil in any of those countries, it doesn't take much to completely pack up and move.*

"And it helps that the Osprey has a refueling probe. We're no longer limited to how far the ship is willing to steam in one day. Now we're limited to how much can that tanker hold?

"And we can put the Marines in the back and tank, and as long as I've got a C-130 that's willing to go with me and has something to give me, it's human limited now. How many hours can I fly this airplane before I'm too fatigued?

Question: The Marines deployed the SP MAGTF in April?

Major Rhobotham: "It was deployed in April and it actually self-deployed. The V-22s flew across the Atlantic, and although it has been done before, this is a new operational reality which folks need to recognize exists. We got all the airplanes where they needed to go flying there, and not being airlifted by the USAF."

Question: In your view how is the SP-MAGTF different from a MEU?

Major Rhobotham: "It compliments a MEU very, very well. It is a different tool set. It is similar to having both a screwdriver and you've got a drill in your toolbox; that drill is a lot like the MEU. It's a lot more powerful, it can go a lot faster; it can do a little bit more powerful things. But it doesn't mean you need to throw away your screwdriver.

"The SP-MAGTF has a lighter footprint, and we can go to any place that the government sees that needs a little bit of attention; we can drop one of these special purpose MAGTFs off.

"We can just go wherever we need to, drop it off, and then when that situation's resolved itself or reached some sort of threshold that we feel comfortable, we can pick this up and move it anywhere we want to.

"In the past we would have to fly in infrastructure or move by ship; establish the infrastructure and the diplomatic agreements to place the infrastructure in country. Now I can fly in the force; stay until I wish or need to depart.

"A special purpose MAGTF is not to replace a MEU; it is to compliment a MEU. And while there are separate commands, they're not led by

the same colonel, they're designed to complement each other, not to replace each other or be lieu of each other. And I think that's probably a point that doesn't get made enough."[6]

FOCUS ON THE PEER FIGHT AND GREAT POWER COMPETITION: 2017-

The Trump Administration re-focused defense on what they referred to as the Great Power Competition. As the 2017 National Security Strategy (NSS) noted:

After being dismissed as a phenomenon of an earlier century, great power competition returned. China and Russia began to reassert their influence regionally and globally.

Today, they are fielding military capabilities designed to deny America access in times of crisis and to contest our ability to operate freely in critical commercial zones during peacetime. In short, they are contesting our geopolitical advantages and trying to change the international order in their favor.[7]

MV-22B Ospreys with Marine Medium Tiltrotor Squadron 262 (Reinforced), 31st Marine Expeditionary Unit (MEU), perform flight operations from amphibious assault ship USS America (LHA 6) in the Philippine Sea, Jan. 30, 2021. Marine Corps photo by Cpl. Brandon Salas.

The Osprey by the time of the 2017 NSS had achieved maturity and was ready to adapt to the strategic shift in U.S. policy. The Osprey was now operating with the F-35B and the USS America had come to the Navy so a new ARG-MEU was clearly being created.

The new con-ops which would emerge with the re-emphasis on great power competition was two-fold: The shift to distributed operations for the joint force and the need to deal with gray zone conflicts posed by authoritarian adversaries. The Osprey was well positioned to contribute to both demand signals.

The nature of the shift and its implications for the USMC was well articulated by the then CO of MAWTS-1, Colonel Gillette, when I interviewed him in Yuma in the Fall of 2020.

Col Gillette: Working through how the USMC can contribute effectively to sea control and sea denial for the joint force is a key challenge. The way I see it, is the question of how to insert force in the Pacific where a key combat capability is to bring assets to bear on the Pacific chessboard.

The long-precision weapons of adversaries are working to expand their reach and shape an opportunity to work multiple ways inside and outside those strike zones to shape the battlespace.

What do we need to do in order to bring our assets inside the red rings, our adversaries are seeking to place on the Pacific chessboard?

How do you bring your chess pieces onto the board in a way that ensures or minimizes both the risk to the force and enhances the probability of a positive outcome for the mission? How do you move assets on the chessboard inside those red rings which allows us to bring capabilities to bear on whatever end state we are trying to achieve?

For the USMC, as the Commandant has highlighted, it is a question of how we can most effectively contribute to the air-maritime fight. For us, a core competence is mobile basing which clearly will play a key part in our contribution, whether projected from afloat or ashore.

What assets need to be on the chess board at the start of any type of escalation? What assets need to be brought to bear and how do you bring them there? I think mobile basing is part of the discussion of how you bring those forces to bear.

How do you bring forces afloat inside the red rings in a responsible way so that you can bring those pieces to the chess board or have them contribute to the overall crisis management objectives? How do we escalate and de-escalate force to support our political objectives?

How do we, either from afloat or ashore, enable the joint force to bring relevant assets to bear on the crisis and then once we establish that force presence, how do we manage it most effectively?

How do we train to be able to do that?

What integration in the training environment is required to be able to achieve such an outcome in an operational setting in a very timely manner?

Question: Ever since the revival of the Bold Alligator exercises, I have focused on how the amphibious fleet can shift form its greyhound bus role to shaping a task force capable of operating in terms of sea denial and sea control. With the new America-class ships in the fleet, this clearly is the case.

How do you view the revamping of the amphibious fleet in terms of providing new for the USMC and the U.S. Navy to deliver sea control and sea denial?

Col Gillette: The traditional approach for the amphibious force is move force to an area of interest. Now we need to look at the entire maritime combat space, and ask how we can contribute to that combat space, and not simply move force from A to B.

I think the first leap is to think of the amphibious task force, as you call it, to become a key as pieces on the chess board. As with any piece, they have strengths and weaknesses.

Some of the weaknesses are clear, such as the need for a common operational picture, a command-and-control suite to where the assets that provide data feeds to a carrier strike group are also incorporated onto L-Class shipping.

We're working on those things right now, in order to bring the situational awareness of those types of ships up to speed with the rest of the Naval fleet.

Question: A key opportunity facing the force is to

reimagine how to use the assets the force has now but working them in new innovative integratable ways or, in other words, rethinking how to use assets that we already have but differently.

How do you view this opportunity?

Col Gillette: We clearly need to focus on the critical gaps which are evident from working a more integrated force. I think that the first step is to reimagine what pieces can be moved around the board for functions that typically in the past haven't been used in the new way.

That's number one. Number two, once you say, "Okay, well I have all these LHA/LHD class shipping and all the LPDs et cetera that go along with the traditional MEU-R, is there a ship that I need to either tether to that MEU-R to give it a critical capability that's autonomous?

Or do I just need to have a way to send data so that they have the same sensing of the environment that they're operating in, using sensors already in the carrier strike group, national assets, Air Force assets et cetera?"

In other words, the ship might not have to be tethered to a narrowly defined task force but you just need to be able to have the information that everybody else does so that you can make tactical or operational decisions to employ that ship to the max extent practical of its capabilities.

There is a significant shift underway. The question we are now posing is: "What capability do I need and can I get it from a sister service that already has something that provides the weapons, the C2 or the ISR that I need?"

I need to know how exploit information which benefits either my situational awareness, my offensive or defensive capability of my organic force. But you don't necessarily need to own it in order to benefit from it.

And I think that when we really start talking about integration, that's probably one of the things that we could realize very quickly is that there are certain, assets and data streams that come from the Air Force or the Navy that make the USMC a more lethal and effective force, and vice versa.

The key question becomes: "How do I get the most decisive information into an LHA/LHD? How do I get it into a marine unit so that they *can benefit from that information and then act more efficiently or lethally when required?*

And with the pairing of the F-35B with the Osprey onboard the large deck amphibious ships the Marines obviously now contributed in a major way to the ability of the Navy to engage more effectively globally dealing with peer adversaries.

But the adaptation of the Osprey itself was underway. During a visit to 2nd Marine Air Wing in 2020, I learned of an exercise in which the Osprey was configured to play a key role as the quarterback of an operation.

I highlighted this exercise and interviewed Major Rew, the exercise's air mission commander in 2020.

U.S. Marines with 2d Marine Air Wing take off after dropping Marines with 2d Battalion, 2d Marine Regiment, 2d Marine Division as part of Exercise Deep Water on Camp Lejeune, North Carolina, July 29, 2020. Deep Water is a 2d Marine regiment-led exercise designed to provide Marine Air-Ground Task Force capabilities, which increases lethality and combat effectiveness for future combat operations. The exercise included the largest air-assault conducted in decades. Credit Photo: USMC, 2nd MAW.

In a press release from November 5, 2020, this is how II Marine Expeditionary Force described the exercise:

Last July, North Carolina-based Marines organized an exercise in which they called Deep Water.

In a press release from November 5, 2020, this is how II Marine Expeditionary Force described the exercise:

"Marines with 2nd Marine Division, 2nd Marine Logistics Group, and 2nd Marine Aircraft Wing are conducting Exercise Deep Water at Marine Corps Base Camp Lejeune, N.C., 29 July 2020.

"II MEF conducts these training events on a consistent basis. This year, Exercise Deep Water will see two battalions conduct an air assault in order to command and control many of the various capabilities organic to II MEF in preparation for major combat operations.

"Exercise Deep Water 20 is a great opportunity for the Division to work with aviation units from Marine Corps Air Station New River and the Logistics Combat Element, as well. 2nd Marine Regiment will be the provide command and control over the 2nd battalion, 2nd regiment, and 3rd battalion, 6th regiment, the logistics and aviation units...."[8]

During my visit to 2nd MAW in the first week of December 2020, I had a chance to discuss the exercise and its focus and importance with Major Rew, the exercise's air mission commander.

I learned from Major Rew that this exercise combined forces from pickup zones in North Carolina and Virginia. The exercise consisted of a force insertion into a contested environment, meaning they used air assets to clear areas for the Assault Force, which included both USMC (AH-1Z, UH-1Y, F/A-18A/C/D, and AV-8B) and USAF aircraft (F-15E and JSTARS). Once air superiority was established, the assault force was inserted by USMC MV-22Bs and CH-53Es. The exercise also included support aircraft such as the KC-130J and RQ-21.

The planning and execution focused on bringing a disaggregated force into an objective area that required integrated C2 with Ground, Aviation, and Logistics Combat Elements. This C2 functionality was delivered in part by an Osprey operating as an airborne command post with a capability delivered by a "roll-on/roll-off" C2 suite, which provided a chat capability and can be found at a mobile or static command post or even in an airborne C2 aircraft.

The use of MAGTF Tablets (MAGTAB) provided a key means of digital interoperability that allowed for real time information sharing to ground elements and aviators. The MAGTAB provided the visual representation of the integrated effects and outcomes to the command element.

ISR was provided by USMC assets and by a USAF JSTARS aircraft. They used their Network-On-The-Move Airborne (NOTM-A) system to provide interoperability for the commander and assault force.

As Major Rew put it:

I think having the NOTM-A kit on the Osprey is a big win because it provides so much situational awareness. With the Osprey as a C2 aircraft, there is added flexibility to land the aircraft close to whatever operational area the commander requires.

There are many capable C2 platforms across the DoD but not all of them also have the ability to immediately land adjacent to the battlefield like the Osprey does.

One aspect of mission rehearsals the Marines are developing is to leverage joint assets in support of an assault mission and be able to provide information to that mission force as well.

To be clear, the Marines did not march to the objective area; they flew to their objectives in various USMC lift assets accompanied by USMC rotary wing and fixed wing combat aircraft. They were moving a significant number of Marines from two different locations, hundreds of miles apart, to nine different landing zones.

Major Rew underscored:

We were working with a lot of different types of aircraft, and one of the challenges is trying to successfully integrate them to meet mission requirements.

As the air mission commander, I was co-located with an infantry colonel who was the overall mission commander. We were in an Osprey for a significant period of time leading the operation from a C2 perspective.[9]

FORCE STRUCTURE RE-SET AND RE-DESIGN: MULTI-DOMAIN OPERATIONS, 2018-

The focus on distributed operations – Maritime Distributed Operations (DO), the USMC Expeditionary Advanced Base Operations (EABOs), and the Air Force agile combat employment or ACE – fits the capabilities of the Osprey – with its speed, range, air refuelable capability – perfectly. It is both an enabler of distributed operations and a benefactor from such operations. It can create distributed points of operation, and can leverage them.

**Multi-mission everything
MV-22B "Osprey"**

Slide presented by LtGen Davis to the Sir Richard Williams Foundation seminar in 2016.

The payload revolution associated with distributed operations is enabling the emergence of a kill web force. But a key to being able to do so is the C2-ISR revolution which can proliferate payloads to be inserted in air, ground, space and maritime assets which can cross-support the operations of a distributed force.

I first witnessed the Osprey being fitted for a new mission with a new payload when I visited the Boeing plant in Philadelphia in 2015 with Murielle Delaporte. Specifically, they were working on the ability of the Osprey to operate as a tanker. And during that visit to the Boeing V-22 Osprey facility, the Boeing team explained the range of possibilities for the Osprey adding multi-mission roll-on-roll off capabilities.

And in 2016, in a presentation in Australia to the Sir Richard Williams Foundation, LtGen "Dog" Davis talked about the transformation of the Osprey into the multi-domain world.

When LtGen Davis, the Deputy Commandant of Aviation, spoke at the Williams Foundation seminar on new approaches to air-land integration, he described a key aspect of the evolving Marine Corps approach with their air assets as "multi-mission everything".

Technology is important to this effort, and he highlighted that the Osprey being brought into the force was a generator of "disruptive change," but the kind crucial to real combat innovation.

He noted: *If we held this conference 12 years ago, and the room was filled with Marines we would hear about all the things the Osprey could not do and why we should not go ahead.*

If we brought those same Marines into the conference room now, they would have amnesia about what they thought then and press me to get more Ospreys and leverage it even more.

But it is not just about technology – it is about "equipping Marines, not manning the equipment.[10]

His point was that you needed to get the new equipment into the hands of the Marines at the earliest possible moment, because the young Marines innovate in ways not anticipated when the senior leadership gets that equipment to them.

As the emphasis on multi-mission capability to enable a multi-domain force accelerates, the Osprey is key enabler of this strategic force structure shift. With the Navy developing the CMV-22B variant of the Osprey, the multi-mission capabilities of the aircraft are clearly being expanded as well.

The CMV-22B provides an important stimulant for the shift in con-ops whereby the Navy's experimentation with distributed operations intersects with the U.S. Air Force's approach to agile combat employment and the Marine Corps' renewed interest in EABOs.

In other words, the reshaping of joint and coalition maritime combat operations is underway which focuses upon distributed task forces capable of delivering enhanced lethality and survivability.

And now with the U.S. Army becoming a stakeholder in the Tiltrotor Enterprise, they will add their own demands on the enterprise but also be able to leverage many years of operational experience by the other stakeholders. And with the emphasis on an open architecture for the Army tiltrotor aircraft, a "mission-kitable" aircraft is assured.

In this 2020 article by Steve Lamb, the Army approach was explained as follows:

Whether performing transport, logistics, strike or reconnaissance duties, helicopters and tilt-rotor aircraft greatly multiply the effectiveness of ground forces. They are a combat multiplier as well as a lifeline in austere environments.

Future Vertical Lift (FVL) will transform the Army's rotary wing fleet, bringing faster, more lethal and more survivable aircraft to the battlefield....

"Future Vertical Lift will be the most sophisticated rotorcraft to enter military service, with all systems connected by the digital backbone," said James Conroy, vice president, navigation, targeting and survivability, Northrop Grumman.

"Just as a mobile phone relies on an operating system to connect apps and sensors, this digital backbone will allow the next generation of avionics and self-protection systems to work in a unified way."

With future generations of FVL aircraft planned to operate alongside the enduring fleet of Apaches, Black Hawks and Chinooks, ensuring interoperability and commonality within Army Aviation will also be a top priority. Investments in FVL must enable the enduring fleet to be capable of multi-domain operations concurrently. [11]

And the Army can leverage the many years of operational experience of the Marine Corps, the Air Force and now the Navy, in operating the tiltrotor aircraft. And the Army will itself provide new energy and concepts of operations to the tiltrotor enterprise as it moves forward into the future.

𝓧 5 𝓧
THE LAUNCH: 2007-2011

I wrote a piece in 2012, which provided an overview on the Osprey at the five year mark and highlighted the initial period of its operation and provides an overview to the launch period.

An excerpt from that article follows:

In September 2007, the Osprey was deployed for the first time to Iraq. In February 2007, a CH-46 was shot down by Infra-Red MANPAD in Iraq. In May 2007, the Commandant and LtGen Castellaw announced the decision to deploy the Osprey as soon as it was ready.

In early July 2007, LtGen Trautman replaced Castellaw as Deputy Commandant of Aviation and the squadron subsequently deployed in late September 2007.

The plane has not only done well, but in five short years has demonstrated its capability to have a significant impact on combat while reshaping thinking about concepts of operations.

At this point, it is important to grasp the lessons learned and shed some light on where the plane and the USN-USMC team might well move into the future. The Osprey provides a solid foundation for innovation and the transformation of concepts of operations for the entire USN-USMC team, assuming boldness will overcome timidity.

Marines prepare to embark an MV-22 Osprey. June 5, 2008, Al Asad, Iraq. Photo by Sgt. Scott McAdam.

In this regard, it is important to go back to the founders of the modern U.S. Navy who understood that strategy is built around what is coming, not where one has been. Along that vein, today's strategy must be built on the capabilities of F-35s and Ospreys, not on F-18s and CH-46s.

As noted in an earlier piece posted on AOL Defense:

"In 1924 a very accomplished Admiral grasped that action/reaction vision for the future. The President of the Naval War College, overseeing these discussions (studying the complexities of British and German Fleet tactics during the Battle of Jutland in WWI), was none other than Admiral William S. Sims, who had already influenced King's and Bill Halsey's development of destroyer techniques, not to mention the convoy system.

"When Sims spread his war games fleet across the plotting board, he introduced aircraft carriers to the mix even though Lexington and Saratoga were still months away from commissioning, and he argued that the aircraft carrier would replace the battleship as the Navy's capital ship.

"The reason was that carriers presented a 360-degree range of fire-power via their aircraft that far outdistanced the radius of a battleships' guns. Sim's fixation with a widening circle of projected power may have influenced Nimitz's fellow classmate, both at Annapolis and now at the

Naval War College, Commander Roscoe C. MacFall, USN when he took his turn at the plotting board. Rather than placing his ships in long lines, MacFall arrayed his fleet in concentric circles around his capital ships, admittedly still battleships.

"The tactical advantage was driven by a common pivot point in the center of the circle where all ships could turn together and remain in formation."

Annually for the past five years, I have interviewed Osprey pilots and logisticians and have had the opportunity to hear firsthand from the operators about what they are doing and what they have learned during this roll-out period.

This year was no different. I spoke with three experienced pilots and leaders in what is called the "Osprey Nation," the growing group of young aviators and maintainers who form the nucleus of the future of the USMC and of power projection.

The decision of USMC leaders to deploy the plane to Iraq was part of the "testing" process.

What is often overlooked is that testing is done by pilots and maintainers in combat, not by technicians in white coats or statisticians at the Government Accounting Office.

There was concern expressed by the Marine Corps Aviators that the deployment to Iraq would prove challenging, and it was. But it was also evidence that leadership was willing to make the hard decision to roll-out needed capabilities and let the users, not the program managers, define the direction of a program.

The deployments have been on land – in Iraq and Afghanistan – as well as at sea. The plane and its crews have been tested in combat and in real world operations.

What we have seen is that the plane started with "training wheels on" its deployments, and those wheels not only have been thrown off, but, as time in combat has gone up, the Marines as well as the Combatant Commanders have begun to understand what this transformational platform can do when connected with other capabilities and assets.

I will never forget a comment made five years ago by a USMC logis-

tician. "The Osprey is not a very good CH-46 replacement." He was right.

U.S. Marine Corps aircraft utilized by Marine Aircraft Group-Afghanistan (MAG-A), sit on the flight line during a dust storm aboard Camp Bastion, Helmand province, Afghanistan, 2014. U.S. Marine Corps photo by Lance Cpl. Darien J. Bjorndal, Marine Expeditionary Brigade Afghanistan.

It was not really a replacement at all, but a step into a very different distributed operations future. The plane demonstrated the speed and range of the aircraft in covering Iraq.

As one Marine commented, "The MV-22 in the area of operation area of operation (AO) was like turning the size of the state of Texas into the size of Rhode Island."

It was the only "helicopter" that could completely cover Iraqi territory. And in this role, the testing of support as well as operational capabilities was somewhat limited as Marines tested out capabilities and dealt with operational challenges. The plane was largely used for passenger and cargo transport in support operations in difficult terrain and operating conditions.

It was used for assault operations from the beginning, but over time the role expanded as the support structure matured, readiness rates grew and airplane availability was increasingly robust. From the beginning, the aircraft impressed and foreshadowed later developments.

As General Walsh, now Deputy Commanding General Marine Corps Combat Development Command, noted in an interview at the Cherry Point Air Station in 2009 after a year in Iraq — with the withdrawal of U.S. forces from Iraq there was a roll-up of forward operating bases. This meant that the remaining forces had to cover more ground and to provide protection at greater distance.

Enter the Osprey, which did not require forward operating bases (FOBs) to provide lift and support to forward deployed forces. General Walsh underscored that as the U.S. forces withdraw there was demand for more, not less, airpower.

"On one level, this was due to the drawdown of the number of combat posts, which supported operations in Iraq. American forces continued to work with Iraqi forces but now had to commute from distance to do their work, rather than being in close proximity to combat posts.

"This meant that airpower had to provide regular support to the transit of U.S. forces working with Iraqis. At one point we had 140 combat posts; while we were there we went from 36 to 4 combat posts; so air was relied on more frequently for convoy protection.

"As we drew down combat posts and associated capabilities, air was relied on for capabilities which had earlier been largely provided by the ground forces. On another level, this was due to the need to protect the convoys moving equipment out of Iraq. As you close down and do retrograde, you have to move further out in road miles and that requires air support.

"In addition, transport needs to move support elements to work with Iraqis increased demands for air transport. We were increasingly asked to provide support for partnering operations."

Iraq was the beginning and a consciousness raiser for troops and commanders.

Next on the agenda was the beginning of deployments to Afghanistan. The Afghan phase of deployments has seen the aircraft and its operator's transition to more assault combat operations over time, to the point where the latest Osprey squadron just came back from Afghanistan with record setting assault operations.

One metric is the number of Named Operations the Osprey squadron participated in in Afghanistan. The Osprey squadrons have significantly increased their involvement in what the military calls Named Operations, and these operations are air assault operations in support of U.S. and coalition forces.

The latest squadron VMM-365 (the Blue Knights) conducted nearly

200 Named Operations, which was a 20-fold increase over the squadron which preceded it in Afghanistan.

Major General Walters during my visit to 2nd Marine Corps Air Wing in early March 2012.

In the words of the head of 2nd Marine Air Wing, Major General Glen Walters, USMC upon his return from Afghanistan:

"The Ospreys had their normal fair share of general support, resupplies, etc. But we started accelerating their use as my time there went on, and used them for both the conventional and Special Forces operations.

"The beauty of the speed of the Osprey is that you can get the Special Operations forces where they need to be and to augment what the conventional forces were doing and thereby take pressure off of the conventional forces.

"And with the SAME assets, you could make multiple trips or make multiple hits, which allowed us to shape what the Taliban was trying to do. The Taliban has a very rudimentary but effective early warning system for counter-air.

"They spaced guys around their area of interest, their headquarters, etc. Then they would call in on cell or satellite phones to chat or track. It was very easy for them to track. They had names for our aircraft, like the CH-53s, which they called "Fat Cows."

"But they did not talk much about the Osprey because it was so quick *and lethal. And because of its speed and range, you did not have to come on the axis that would expect. You could go around, or behind them and then zip in. We also started expanding our night operations with the Osprey. We rigged up a V-22 for battlefield illumination.*

"A lot of these mission sets were never designed into the V-22, but you can put it into the field and configure it to do various missions, as required. We have new software for the Ospreys in Afghanistan where you can pick your approach, angle, approach speed and let the aircraft do it all. That is a huge safety gain."

The start of this transition to a tip of the spear air assault capability was seen at the beginning of the Osprey's deployment in Afghanistan.

According to LtCol Bianca, the Osprey squadron commander at the time:

"It is one thing for me to do an assault support mission where I insert troops to a location. It is quite another to talk about distributed operations... From the distributed angle, never forget that the troops are not here at Camp Leatherneck; they are always somewhere else... Your only option is to get into a V-22, because 'I got to get to that corner in the open world — no roads, nothing there — we got to go do it', and that, then, becomes our mission."

THE EVOLUTION OF THE OSPREY: "WE ARE NO LONGER A BAR ACT"

September 9, 2012

Another good look back at the launch period was provided in an interview which I did in 2012, which was the five year operational benchmark for the Osprey. In a discussion with LtCol McAvoy, the CO of VMM-264 at USMC Air Station New River, he laid out an overview of the progress which had been made over the first five years.

LtCol McAvoy was a CH-46 pilot for nine years and then transitioned to the V-22 in 2004. He went on the third Osprey deployment to Iraq in 2009 and then to Afghanistan for the third deployment there in 2011.

LtCol McAvoy provided insight into the differences in how the Marine Corps was able to approach the Osprey and its use in the two deployments as well as to provide insight from flights to participate in the Farnborough Air Show in 2006 and then again in 2012.

In a fundamental way, LtCol McAvoy and his experience sheds light on the evolution of the Osprey as a weapon system used by the USMC in operations, both non-combat and combat.

Question: What were some of the major differences between the flights made in 2006 and 2012 with the Osprey to participate in the Farnborough Air Show?

LtCol McAvoy: *In 2006, the flight to Farnborough was the first transatlantic flight of the Osprey. We flew 2 Ospreys to Farnborough and it was challenging.*

One of the aircraft made it all the way to the U.K. without an issue. The other V-22 experienced "an engine compressor stall" and was forced to make a precautionary landing at a U.S. military base in Iceland. We accomplished the trans-Atlantic mission, but much of what we did was "a first," so we were learning along the way.

Fast forward to 2012, with multiple deployments under our belts and over a hundred thousand hours behind the Osprey Community, the transatlantic flight was a totally different experience.

We were required to have 4 Ospreys in Farnborough, so we started out with 6 flying from New River to St. John's Newfoundland, where we stepped off. The aircraft were all configured with three internal fuel tanks to give us about six hours of fuel. This has become standard operating procedure for long-range flights.

In St John's, we were joined by three KC-130Js; one tanker, one trail maintenance and one back-up. The next morning all aircraft took off without an issue. One hour into the flight, the back-up aircraft (2 MV-22s and 1 KC-130J) returned to St John's.

The rest of the package took approximately 5,000 lbs of fuel and continued on the 5.5 hour leg to Lajes, Azores. We stayed overnight in Lajes, and the next morning we again departed without issue. Four Ospreys and two KC-130Js flew directly to Farnborough from Lajes,

Azores in about 6 hours. We had no significant maintenance issues along the way.

LtCol McAvoy after the interview.

Question: How would describe the difference from 2006 to 2012 at RIAT and Farnborough?

LtCol McAvoy: *In 2006, it felt like we were a bar act. It was challenging to get there and we were seen as oddities. In 2012, we were flying a plane with years of combat experience. We no longer were a bar act, but war fighters flying and maintaining a key combat capability.*

We had a story to tell. We had folks with hundreds of hours of operational experience behind us. It was a proud moment for me as the CO. I was the senior guy and I had no majors; it was all captains, all young guys who really shined in explaining the aircraft to others.

In 2012, it was the young, but very experienced enlisted Marines and officers who told the MV-22 story, and their stories were based on actual combat.

Question: Let us focus now on the differences from the 3rd deployment to Iraq and the 3rd deployment to Afghanistan, a gap of only two years. How would you describe the differences?

LtCol McAvoy: *We were focusing on the basics in Iraq. In my squadron at the time, no one had deployed in combat with this aircraft. Some had deployed with other type model series, but this was the first with the V-22.*

In effect, Iraq formed a foundation for follow-on operations and we just gained experience from that deployment and minimized risks in later deployments. We grew together as a team and built a solid foundation during that Iraq deployment.

When we went to Afghanistan last year, we leveraged that foundation.

We were familiar with the aircraft in tough operational conditions. We were able to expand the box exponentially. We moved on from logistic support to doing Named Operations.

We have become a high performance assault platform. We are doing large movements day and night into the worst zone, hot zones, places where we know that we are going to have contact with the enemy.

In other words, as the experience grows from the team, we can expand what we can do with the aircraft. And the young guys with just V-22 experience are pushing the envelope.

During our time in Iraq, we used largely helicopter tactics. In Afghanistan, we were using V-22 tactics. The military as a whole is now willing to exploit the V-22 capabilities and not just treat it like it was a helicopter.

THE 22ND MEU ACE COMMANDER ON OSPREYS, HARRIERS AND THE FUTURE

August 7, 2013

Another good look back at the launch period was provided in an interview which I did in 2013, with LtCol Schoofield, the ACE commander of the 22nd MEU. In a visit to USMC Air Station New River, I talked with him about his assessment of the Osprey and looked back at the previous years.

LtCol Schoofield is a rarity; he has piloted both Harriers and Ospreys in combat, which makes him the Marine's Marine aviator. He discussed with us the similarities and differences in flying the aircraft and their missions, as well as preparing for the upcoming MEU deployment.

Question: You are both an experienced Osprey and Harrier pilot. Could you comment on the two aircraft and how they are similar and different?

LtCol Schoofield: *I have more than 1000 hours of total time in the AV-8, and have flown in Iraqi (OIF 1 and OIF 2) and Afghani missions. I later transitioned to the MV-22 Osprey and flew the Osprey in Afghanistan for almost a year.*

As far as flying the airplanes they're strikingly similar. They have more in common than you might think. They're two of the three" powered-lift" aircraft ever built, the Osprey and the V-22.

Their ability to take off and land within the lateral confines of their own dimensions then cruise on wing-borne flight is their most unique feature.

In the case of both of those aircraft, you have to carefully manage your gross weight which you can only do by jettisoning fuel so that you maximize your performance at your landing point.

In the Harrier that comes in the form of having your maximum "bring back", so coming back and landing with an X amount of fuel and ordnance onboard if you can. In the Osprey, it comes in the form of landing on the objective with the maximum number of troops and cargo to conduct the mission which has been assigned and with maximum fuel to do follow-up operations.

The two are very similar philosophically in how you think through the mission sets.

Question: How do differences in mission sets affect the approach to operating the planes?

LtCol Schoofield: *The big distinction is the difference between offensive air support, and the assault support mission.*

There is the excitement of flying a single-seat fighter jet in combat, and I can remember flying over Baghdad in 2003 like it was yesterday, and how I was challenged in those scenarios.

But those missions, even the most complex close air support missions that I flew similar to others in my generation did and dropped the ordnance dozens of times in combat, again, not special. Everyone from my generation did.

Those missions are all pretty scripted, even in the most complex missions. They were rigidly scripted as far as what time you took off, what time you landed, what ordnance you had, and what procedures you used.

The assault-support mission is virtually the polar opposite of that. It involves an extreme amount of flexibility. It's a very dynamic mission. And when you conduct assault-support, they're very long missions.

A really long day of flying the Harrier might be five or six hours. A long day of flying the Osprey is 12 or 14 hours actually strapped in into the aircraft. And while you're flying the Osprey, or any assault-support mission, it is a long day.

In a Harrier, you have just one other captain or major that you have to communicate your plan with and again it is in a much scripted way.

With the assault support mission you are dealing with 4 other people in your plane and 4 others in the second ship, so there are 7 people you are communicating your plans with all the way to completion.

I would say, more often than not, you show up to pick up your supported unit, your customers, or the ground force that you're inserting, there's always some subtle changes in play.

You're expecting 12 people and you have 18. Or you're expecting 18, and you got 12 and they forgot to include the dogs or the interpreters or the extra pallet of ammo or fuel or whatever it is.

Things are always different. And anyone who's been flying assault-support for decades could explain that in painstaking detail.

But with that, comes a lot of challenge, which I enjoyed.

Sorting through the mathematics of updating the plan on the fly and communicating that plan to the crew in order to get the maximum number of troops on time at the right place, and coordinating with fires and doing it safely is a core challenge. It's a wonderful challenge and a privilege to experience.

Question: Given your duality of flight experience, you must be looking forward to your assignment as an ACE Commander?

LtCol Schoofield: *I feel very well prepared and we'll have an opportunity to really achieve some great results. The only thing I would like to change, if I could, would be to be current in the Harrier so I could fly the AV-8 and the V-22 and supervise, mentor and lead both my offensive air support and my assault-support pilots at the same time like a carrier air group command would.*

Question: What does the Osprey bring with its speed and range that the helicopter does not for your role as an ACE commander?

LtCol Schoofield: *As you know, there were many critics of the Osprey. This was from people who refused to accept the emerging technology of the Osprey. These were the grandchildren of the critics of the jet engine or of the original helicopters – it cannot work because it did not exist before, or from people with vested interests in maintaining helicopter technology as the core baseline for the force.*

The difference in speed is dramatic. I remember flying up the Potomac River on a V-22, literally passing a Black Hawk like it was standing still. They were flying the helo route from Quantico up to D.C. as were we, but they were going 130 or 140 and we were going closer to 300 knots.

It's a good-looking little helicopter but as I passed that Black Hawk, I asked myself why are we moving people in Black Hawks still?

The United States Army is moving people with these helicopters. Each one has a crew of three or four and carries 12 soldiers slowly to the battle.

Meanwhile, we are moving with much greater speed and safety and with significant operational flexibly and carrying 24 Marines with our 3-4 man crew and going at 300 knots ground speed.

It is the difference between walking and driving. This is a big difference because combat operations are about an ability to put force on target before the enemy moves or is more effective.

The Osprey is part of building that advantage.

A LOOK BACK AT THE LAUNCH PERIOD BY LTGEN (RETIRED) TRAUTMAN

In an October 2020 interview I did with LtGen (Retired) George Trautman, the DCA who sent the Osprey to its first deployments, we discussed the launch period. That discussion provides a very insightful look at this initial introductory period.

Lt. Gen. George J. Trautman III (right), deputy commandant for aviation for the Marine Corps, walks with Col. Kevin S. Vest, commanding officer of Marine Aircraft Group 40, Marine Expeditionary Brigade-Afghanistan, following their flight on an MV-22 Osprey from Marine Medium Tiltrotor Squadron 261 from Kandahar Airfield, Kandahar province, Nov. 24. Trautman spent time in Kandahar visiting the MAG-40 squadrons stationed there prior to his visit to Camp Leatherneck to complete his tour of the entire MEB-Afghanistan aviation combat element. Camp Leatherneck, Afghanistan, November 24, 2009. Photo by Staff Sgt Roman Yurek.

LtGen Trautman: *Many people probably forget where we were in Iraq in 2007. We were in a pretty tough fight. We had gone in there thinking that it would be a cake walk because that's what the administration wanted people to believe. It turned out not to be that.*

We were in a very difficult fight. The Marine Corps Commandant, General Conway, wanting to get what he considered to be the best of the best into that fight, decided in the Spring of 2007 that we would deploy the V-22 to Iraq.

When I arrived in June of that year to take over as the Deputy Commandant for Aviation, the Marine Corps and the Navy support functions, Naval Supply System, and Naval Air Systems Command were struggling to meet this challenge prior to the program's upcoming material support date. There was also resistance down in II MEF and 2nd MAW to making this deployment in the fall of 2007.

The aircraft readiness and supply systems were not where you wanted them to be. The squadron maintenance departments were struggling to move from the analog CH-46 to the digital V-22. Readiness and reliability were not where they would normally be for a typical squadron deployment into combat.

But we could not leave the V-22 out of the biggest fight the Marine Corps has had since Vietnam. We owed it to our Marines to get it in the fight because our troops were at risk. It was the right thing to do.

One of the things we had to decide up front was how we were going to get them there. Today we fly long, long distances with our Ospreys, but in those early days, we still had some issues with the air refueling probe. There were some challenges associated with air flow around the probe and some other minor anomalies.

To transit the Atlantic Ocean, we would have needed to stop three times without aerial refueling so I made the decision that we were going to take them by ship.

Now, going by ship is a typical way for the Marine Corps to send its squadrons all over the globe. It wasn't an atypical thing to do. It was very normal. So, that's what we did.

We put them on a ship and sailed them off the coast of the gulf and flew them into Iraq. In that first tranche of trying to make sure we minimized risk, we made the decision to send them on the ships.

The other risk factor we dealt with was range and depth of supply. Logistics is always a key limiting factor with any global deployment, but it can present a particular challenge when deploying an aircraft type for the first time.

One of the things we did is we decided to overstock the squadron's supply pack up. We built a stockpile of supplies that was bigger than a Marine unit would typically take on a deployment.

Because we are an expeditionary force, we don't typically do that, but the increased range and depth of supply enabled us to maximize aircraft readiness and we made three highly successful back-to-back squadron deployments to Iraq, before later sending additional Osprey squadrons into Afghanistan.

Trautman noted: *The fight in Afghanistan and the fight in Iraq were very different environments. In Afghanistan, you didn't have a road network.*

It also had rugged terrain with pockets of determined foes who you could maneuver and attack from a different direction. That was a very effective operating environment for the Osprey - perhaps even more so than Iraq.

6

SEA LEGS: 2011-2015

The USMC leadership had planned from the outset to operate the Osprey from the seabase. In fact, the Navy had modified an amphibious ship to start such operations.

But the priority of DoD was on the land wars, and so that is where the Osprey went initially. But as the land wars continued, where the focus of activity for U.S. forces was primarily located, the Marines and Navy began the process of exploring what the Osprey brought to the sea-based force.

This became evident in two obvious ways.

First, the Navy and the USMC focused on a series of high-profile expeditionary warfare exercises. Second, the NATO action against Libya in 2011 was a "lead from behind" exercise for the Obama Administration.

No large deck Navy carrier was committed to the sea strikes on Libya only an amphib but the presence of the Osprey turned into a page one story when it became part of ensuring that a captured American pilot was not the story. The Marines turned "lead from behind" on its head.

BOLD ALLLIGATOR EXERCISES

Our team of analysts attended several of the Bold Alligator series exercises from 2011-2014 and witnessed how the USMC, the U.S. Navy and a number of allies and partners joined in the relaunch in many ways of the next generation of amphibious warfare. For that relaunch was happening, as the Osprey was joining the fleet.

The launch of Bold Alligator was in 2011 and this the statement the Public Affairs Office of Expeditionary Strike Group Two and Commander of 2nd Marine Expeditionary Brigade issued prior to the first exercise:

Commander, Expeditionary Strike Group Two (ESG 2), and Commander, 2nd Marine Expeditionary Brigade (2nd MEB) in coordination with ships assigned to the U.S. Second Fleet will conduct a joint large-scale fleet synthetic training amphibious exercise, which will revitalize the fundamental roles as "fighters from the sea."

The exercise, Bold Alligator 2011, is a U.S. naval amphibious exercise that will focus on conducting major amphibious operations simultaneously with a non-combatant evacuation. It will be the first ESG/Marine Expeditionary Brigade-level exercise in almost 10 years.

The exercise will focus on the fundamental aspects and roles of amphibious operations to improve amphibious force readiness and proficiency for executing the six core capabilities of the Maritime Strategy –

forward presence, deterrence, sea control, power projection, maritime security and humanitarian assistance/disaster response.[1]

Bold Alligator 2012 (BA-12) was a training exercise for the expeditionary strike group, and the shaping of a new template for the amphibious task force with the coming of the Osprey and the anticipated arrival of the F-35. The template introduced in BA-12 provided a lay down within which force modernization associated with the F-35B and the VM-22 unfolded.

What was evident in that exercise, and the Bold Alligator exercises which followed, was that the amphibious fleet was shifting from using the ships to function largely as a Greyhound bus carrying Marines ashore to becoming a strike force at sea, and from the sea. At that 2012 exercise we discussed the effort being generated at the time to lay down a new template with officers involved with that exercise.

Capt. Sam Howard at the time of the exercise was Special Assistant to the Chief of Staff of U.S. Fleet Forces Command in Norfolk VA. Marine Col Phil Ridderhof, senior Marine Corps adviser to U.S. Fleet Forces Command in Norfolk, Va.

Capt. Howard began his career in destroyer operations and most recently before his Norfolk Assignment was the skipper of the USS Bataan. During his time on Bataan, he participated in the Haiti relief efforts.

Capt. Howard emphasized that one of the opportunities generated by the exercise was to familiarize other services with the key advantages provided to the joint force by seabasing.

He underscored: *A seabase is a concept that regrettably is foreign to the other services, at least in real practice; and we were able to certainly demonstrate it in real practice in real-time in the case of Haiti and will be able to expand on in the exercise.*

Howard noted that the importance of the exercise as a combined arms approach and was useful to re-shaping military doctrine.

It's a combined arms endeavor, and so getting to that thought process of making it a combined arms thing is certainly very important to us.

Col Ridderhof picked up on the combined arms theme introduced and discussed by Capt. Howard. He emphasized that the exercise was built in part around operating at a different level than an Amphibious Ready Group-Marine Expeditionary Unit (ARG-MEU) was capable of operating.

He argued: *The ARG and the MEU are very important, very capable. When you start getting bigger than the ARG and the MEU an amphibious operation is not simply a quantitative increase. There's a qualitative difference in how you think about organizing the task force.*

The ARG-MEU is built around three core ships and the MEU is of a regimental size, with a battalion landing team, combat logistics, battalion, and the composite squadron. The MEB being used in this exercise is a different animal.

One of our challenges in the Marine Corp is how do you describe the MEB? It can be anywhere from 10,000 to 17,000 soldiers, but in general, it is a regimental landing team with at least three battalions, plus a Marine air group size of an aviation combat element and combat logistics regiment-sized support element.

One of the key things is that to do all six functions of Marine Corp Aviation, to do command and control, is central to the operational capabilities.

A MEU has little pieces, but to do the big Marine Corp Aviation Command and Control pieces is a significant jump in capability. You need all of the pieces of Marine aviation, and then all the logistics to sustain that for 30 days plus.

And the large-deck carrier is shifting its approach as part of this operational construct.

Col. Ridderhof underscored: *Bold Alligator is testing today's capability but because we haven't done this this way, the Combined Force Maritime Component Commander (CFMCC) has typically not been considering the littoral as a whole, going all the way to that objective ashore as his battle space.*

He's influencing it surely but hasn't been thinking of it all the way there. Just as the ARG MEU would be first on the scene, how does the Carrier Strike group fold into the scalability of the operation?"

Capt. Howard emphasized: *The reality is, all this is in the context of a joint war fighting organization. And among the lessons we will take from this is how to develop further the unique war fighting relationship between Navy and Marine Corp and how does that best plug into the total force, joint force.*

Obviously, the refocus on the strategic shift from the primacy of the land wars to engaging peer competitors was already underway in this Bold Alligator 2012 exercise.

BGen Owens after the 2012 interview.

In our interview with BGen Owens, the II MEF Commander, held after that exercise, he highlighted how he saw the exercise effort.

One of the things that was different in this exercise from many previous amphibious large-scale exercises is we executed in what we called a medium threat, anti-access, area denial (A2AD) environment. The threat focus is primarily on the area denial piece, which is closer in, but which is more realistic for the timeframe of the exercise.

The threat we faced at sea started with submarines, missile patrol boats, fast-attack craft, fast inshore attack craft, and some asymmetric threats with commandeered fishing boats, low slow-flyers, and some tactical air. But of greater concern was coastal defense cruise missiles, initially fixed sites, as well as mobile, and then ultimately, just a threat of additional mobile sites.

And then, the most ubiquitous threat that we're going to face is mines. In the exercise, we faced a very robust mine capability. We had a wide range of capabilities on the Navy side to help deal with those threats, but we also integrated the MEB in that, particularly our air assets.

These assets were used both in targeting threats to the amphibious task force ashore, as well as providing defense of the amphibious task force primarily with our aviation asset. But we also involved some of our ground combat elements when they were aboard the ship.

That continued even after we went ashore. And this is something that we really haven't practiced; this full integration, of the Marine capability in the overall ability to both to project force, and to protect those naval assets that are projecting that force.

The exercise was focused on redesign of the current force structure to achieve the desired combat effect in order to lay down a template for the changes to come.

As BGen Owens put it: *I think flexibility is a key word. In this exercise, we focused on today's forces for today's fight. What it really was about was getting the greatest impact, the greatest benefit out of the capabilities we have.*

For example, in our countermine effort, we recognize that in order to do countermine work here on the east coast of the U.S., it's going to involve coalition participation. So, we had Canadian mine hunters out working with U.S. divers in conjunction with Dutch divers, Canadian divers supported by the Coast Guard providing a cutter to help provide force protection for the mine hunters.

And our Navy forces provided close in protection for both the mine hunters, and then subsequently, for some of our maritime sealift command shipping that was coming into the same areas.

Thus, in addition to integrating the Marine and Navy pieces, we also expanded our search to what other capabilities that other countries, other services, for instance, the coast guard, and even the interagency could provide.

We didn't really touch on the inter-agency aspect too much in Bold

Alligator 12, but it is an aspiration for the future. We want to be able to tap into capabilities that will help us defeat some of these asymmetric threats, in particular, in order to project force ashore.

BGen Owens underscored a very key point about shaping a way ahead with the force you have to shaping the future force.

From this point of view, the goal of the exercise was to shape an effective concept of operations with current capabilities. We've got to have the concept of operations in place as we integrate new capabilities going forward.

The latest capability which was included at the time was the Osprey. In fact, while journalists waited on shore for the insertion force, the Osprey team had led an assault deep into the battlespace. And even more interestingly, the Ospreys launched from a range of sea-bases, including a Military Sealift Command ship.

And with the launch from an MSC ship, the Osprey was highlighting the next phase, rather than focusing on the ARG-MEU, one could think in terms of an amphibious task force.

This is how BGen Owens highlighted the importance of the event.

The T-AKE is bringing in our dry cargo. So they bring in beyond what the amphibious ships carry, they'll bring in food, water, and they are ships that bring in our ammunition. That was what we exercised using the V-22s to land on the T-AKE, and we had our logistics regiment Marines posted aboard the T-AKE to work on the distribution piece.

In fact, the "return to the sea" energized by the training efforts in the Bold Alligator exercises came at the same time that the Osprey was making its broader impact on the USMC. There is no more dramatic case of a platform introducing disruptive change in a service than the Osprey to the USMC.

But it took a while for the concepts of operations to change and for commanders to understand fully that they did not have to operate in a constricted operational box of a few hundred miles for the ARG-MEU.

Now they could think about a 1,000 mile operational area at

sea. As one Marine at the time of the Bold Alligator exercises in the period described the transition and the challenge to adapting to what the Osprey can do with fast jets:

The speed and range of the aircraft is a game changer. But it's also the endurance of the aircraft. Once it's flying, it's flying. And we had a lot of missions that required flight time more than six hours, which is very taxing for the jet guys, and for us it is as well, though not as bad because we can trade off in the cockpit.

The fact is that you can have airborne assets — both as a package as well as a Tactical Recovery of Aircraft and Personnel (TRAP) for sensitive site exploitations — airborne all at the same time for hours at a time to respond to something that happens in the AOR. It allows maximum flexibility for response time down to something like thirty minutes, depending on where it is.

And then you can sanitize the scene from there and everybody returns home. It's a capability that hasn't been overlooked, but hasn't been utilized like that. We just didn't really have that capability before, especially on much longer ranges and in short response time.

So by marrying those two with the fixed-wing aviation asset we can do operations differently. We can neutralize a target and then immediately have a strike team insert to confirm that whatever happened really happened, and then deliver whatever materials they need, get back on an aircraft and leave in under thirty minutes in any location that we're operating on a 600-mile ring. This is amazing!

THE LIBYAN OPERATION, 2011

Enter Libya, and linking the Osprey to the USN-USMC "Gator" Navy opened up a whole new capability. The "Gator" Navy began its transition from Greyhound Bus to a new strike force capability.

The ability to link seamlessly support services ashore with the deployed fleet via the Osprey allowed the Harriers aboard the USS Kearsarge to increase dramatically their sortie rates.

The future was being redefined by the Osprey.

LtCol Boniface during the 2011 interview.

LtCol Christopher Boniface, USMC, commanding officer of VMM-266 and the Osprey leader in Operation Odyssey Dawn, highlighted in an interview I did with him in New River in 2011 the sense of the change involved in the coming of the Osprey to the USMC-U.S. Navy team.

Question: Some folks consider the Osprey as the replacement for the CH-46 much like some folks consider the F-35B as a replacement for the Harrier.

But I believe that this actually distorts the discussion because bringing both planes to the MEU is a game changer. What are your thoughts about the process of transition?

LtCol Boniface: *The Osprey is clearly not a CH-46. It is an aircraft that can fly like an airplane, but land and takeoff like a helicopter, but it is not defined by its essential ability to operate as a rotorcraft.*

In comparison to the legacy CH-46E, I can carry twice as much, twice as fast and twice as long. Actually, in many cases I can almost triple the capability. Finally, it is a more reliable aircraft and ultimately a safer aircraft.

It provides the ability for the PHIBRON/MEU to keep the ARG farther out to sea if needed, thus increasing the element of surprise and keeping us safer too. With the CH-46E, you are typically operating 25-50 nautical miles (NM) from shore. As of today, I can operate 250 NM or greater from shore and I can close this 250 NMs from ARG shipping in just about an hour with 7,000 pounds of Marines, or cargo in the back.

I can actually launch at approximately 52,000 pounds due to the MV-22's ability to perform a rolling short takeoff (STO) from the flight deck as opposed to the vertical takeoff required by a traditional helicopter.

Bottom line, I can carry more Marines, cargo and equipment, and I can close an objective area twice as fast while staying outside of most enemy weapon engagement zones.

It is a completely different animal.

It is a true "game changer."

Comment: The Osprey is significant in logistics support for the fleet as well which was demonstrated during recent operations off of Libya.

LtCol Boniface: *I need to be very clear, the Osprey is not only supporting the USMC, but also supporting the USN-USMC team. During this deployment the USS Kearsarge suffered a mechanical loss of a propulsion screw.*

We were only able to do four knots through the water; at that point we were 300 miles from land. The only thing we could do was to get tech reps and parts out to the ship to allow us to make a best speed of 11 knots to get back into the fight.

Remember you can't really launch aircraft if you can't make the correct wind across a flight deck, and you can't effectively launch Harriers to continue their strike mission with four knots of wind either.

The Osprey was the only bird we could use to close this gap, fix the ship and continue the mission we were executing off the coast of Libya.

The V22 is like driving a Cadillac Escalade compared to your dad's old truck. I'm traveling better; I'm able to carry more It's more reliable, and efficient. It's smoother. It's safer.

Really comparing the CH-46E to the Osprey is like comparing apples and oranges. With the CH-46E, I am typically flying 300 feet at 110 knots. With the Osprey, I can be at 13,000 feet and flying at 250 knots, all with Marines and equipment in the back. I am flying an airplane, not a helicopter.

Question: When we were talking earlier, you empha-sized that the Osprey's capabilities compared to the Battle Phrog was game changing in character. Could you elaborate?

LtCol Boniface: *It completely changes the game for the ARG/MEU, it changes the game for how the Marine Corps does business. I didn't fully realize, nor appreciate this until I was operating in some of these locations during our deployment.*

Once we got into the Med for the Libyan operations during Operation ODESSEY DAWN, Naval Air Station Sigonella was our only forward support base.

The Osprey functioned as a force multiplier in these circumstances.

I could fly 300 miles plus from the USS Kearsarge to Naval Air Station Sigonella, land, get a quick hit of gas if needed, put five, six, seven thousand pounds of gear, equipment, troops, parts, and be back quickly to the ship within 2.5 hours.

Half of our MV-22s were conducting combat operations in Afghanistan while we were conducting combat operations off the coast of Libya aboard the USS Kearsarge.

So you can do the math: Half of the Osprey's conducting combat operation in Afghanistan and the other half performing combat resupply, and TRAP operations off the coast of Libya.

I wouldn't have even fathomed this expeditionary and amphibious capability 10 years ago. Also, the Ospreys from Afghanistan flew directly to Souda Bay, Crete and then onto Naval Air Station Signalla, Italy.

This trip is a 3500 NM transit. This has been the longest in our short history, and they did it in one day. You can't even begin to argue or compare and contrast these facts with the CH-46E.

Question: How important to the operation was it to break the CH-46 tether?

LtCol Boniface: *A complete transformation to how we are doing business has been involved. In order for the USS Kearsarge, the ARG and the 26th MEU to stay in their operational box during Operation ODESSEY DAWN, and enable the Harriers to continue their strike mission, we were reliant on other assets to supply us.*

For many supply items, the Osprey provided the logistical link to allow the ARG to stay on station and not have to move towards at sea re-supply points and meet re-supply ships.

Without the Osprey you would have to pull the USS Kearsarge out of its operational box and send it somewhere where it can get close enough to land or get close enough to resupply ships to actually do the replenishment at sea.

Or you would be forced to remain where you are at and increase the time you're going to wait for this part by three, four days or even a week.

The ARG ships are only moving at 14-15 knots. At best, let's just say they move an average of 13 knots per hour, and add that up for the 300 miles that you have to sail.

Now you're looking at least a day to get the needed folks, parts or equipment and then the transit time back to the operational box. The V22 will do that in a couple hours and allow the ARG/MEU to keep executing its mission.

Question: So part of your game changing argument is something people do not usually focus on, the use of the Osprey as a backbone of rapid resupply and logistics support?

LtCol Boniface: *It was a key performance element. It facilitated the ARG's strike missions over Libya by allowing the Osprey to perform a combat resupply mission.*

And it did the other items crucial to the operations such as the TRAP mission and bringing in supplies to keep the Harriers operating. By keeping us within a suitable striking distance, the Osprey helped reduce the AV8-B pilots overall fatigue factor, increase that safety margin and ultimately kill bad guys.

And if the collapse had been more rapid, and we would have needed to put Marines on the ground (for example, a possible humanitarian assistance, disaster relief mission), there is no question that the Osprey would have been central to that effort as well.

7

FOCUS ON PACIFIC
OPERATIONS: THE PIVOT

The Obama Administration highlighted what they called "a pivot to the Pacific." The idea was to drawdown engagement in the Middle East and re-focus on Asia. As Congressional Research Report characterized the effort:

In the fall of 2011, the Obama Administration issued a series of announcements indicating that the United States would be expanding and intensifying its already significant role in the Asia-Pacific, particularly in the southern part of the region. The fundamental goal underpinning the shift is to devote more effort to influencing the development of the Asia-Pacific's norms and rules, particularly as China emerges as an ever-more influential regional power.[1]

But this"pivot" or "rebalancing" toward the Asia-Pacific was supposed to happen aa time of fiscal constraint and what would be an expansion of the engagement in Afghanistan, this meant that very few new resources were committed to the effort.

The Marines were also asked to engage in force re-positioning in the region, which with no new resources was more than a modest challenge. And the coming of the Osprey was a crucial element of being able to start the re-positioning effort with the limited resources which the Marines had available.

When one looks back at the very negative press the Osprey had at the time of its introduction in 2007, it is amazing to see it emerge as the key lynchpin of the Marine Corps contribution to the "pivot" to the Pacific.

One could write a book about this period. Well actually I did with two colleagues. We published a book in 2013 entitled *Rebuilding American Military Power in the Pacific: A 21st Century Strategy*. And on the cover of that book we had a picture of the Osprey landing on a large deck carrier which would presage the acquisition by the Navy of the plane in the 21st century.

In this chapter, I will start with our perspective written in 2013 about the role of the Marines in the pivot to the Pacific and how the Osprey fitted in to that strategy. In the companion volume to this book. I have provided additional interviews conducted at the time with the MARFORPAC team with regard to how they were approaching the challenge and the role of the Osprey in facilitating their role.

THE PERSPECTIVE FROM 2013

Chapter twelve in our book on rebuilding American military power in the Pacific specifically looked at the USMC role in this effort. We argued the following:

The USMC as the Spearhead of Change

The building blocks for a new approach are being lived right now by one of the core forces making up the U.S. Armed Forces. The USMC is in the pivot to the pivot to the Pacific and are deploying a key asset that is a distributed operations game changer — the Osprey — and are preparing deploy another — the F-35B.

Of course, the other key elements of the U.S. services — the USN, the U.S. Army, the USAF, and the USCG — are key players in shaping the evolution of strategy and capabilities as well.

The point here is more direct — none of these services has bet its future more on the new approach than has the USMC. As a result, as the USMC shifts from the land wars, it is forced to draw the lessons learned from this decade of experiences and apply them to the next.

And the USMC is basing its future on the Osprey, the F-35B, and the new ships it is operating from with the USN in extended littoral operations and flexible engagements.

As the lead force deploying the F-35, they are in unique position as well to work with the USAF in reshaping a global approach to the role of airpower as well.

Former secretary of the USAF, Mike Wynne, in many ways the Billy Mitchell of our era, has underscored the key role that the Marines are playing in shaping the new approach.

Wynne is a graduate of West Point, intimately involved in the development of the M-1 tank and the F-16 at General Dynamics and the father of the introduction of Rover, which is the key system linking air systems to the ground via video and many other innovations to his credit or to the teams that he has led. If there is any contemporary American who understands the art and challenges of transition, it is Mike Wynne.

Wynne argued: *It is noteworthy that to truly exploit the significant discontinuity that the fifth generation affords will require experimentation to determine the most effective operational approaches. Such experimentation has already commenced within the U.S. Marine Corps. They see the formation of a very different version of the Army Air Assault force introduced in the Ashau Valley and documented in When We Were Soldiers and Young.*

This updates the MEU in a decisive way and portends potentially the era of independent action with an Expeditionary Assault Group. Experimentation can and should as well be with the fleet, and combine fluid command between the ship and the shore with directed indirect ships fire in support of the dynamic air assault.

The utility of the 360-degree sensors and the utility of the z-axis providing for vertical assault and support are being fully tested by the Marines.

As in the case of the ROVER link, the testing is in the hands of the operator and the feedback is direct and forceful. Whereas there are restrictions between the test community and the Corporate Engineering staff, no such restriction occurs when trying to make or extend the mission or mission set in real operations.

What the USMC has set in motion, the USAF needs to expand upon. Similar to a children's game of leap frog, the USAF and the USMC can tag team to drive con-ops innovation.

The Air Force F-35s needs the same pressure of experimentation and operator feedback as the Marines have set in motion....

The USMC is a key lynchpin in the pivot to the Pacific. USMC forces in Okinawa are moving partly to Guam, and the Marines are shaping a new working relationship with the Australians in Northern Australia.

In fact, they are the lead force in reshaping presence in the Pacific over the next few years. The Marine Corps in the Pacific faces a myriad of challenges. The Marines have been directed through international agreements, spanning two different U.S.

administrations to execute force-positioning moves. This is political, but it is not partisan.

The U.S. Secretary of Defense has mandated that at least 22,000 Marines in PACOM remain west of the international dateline in the distributed Marine Air Ground Task Force (MAGTF) laydown, and he, Congress, and the American people are not interested in a nonfunctional concept for a USMC force.

And the Obama White House has directed the USMC to shift as well forces from Okinawa to Guam and to shape a new working relationship with the Australians.

Beyond what is directed, the Marines need to maintain a ready-force in the face of existing training area encroachments, plus they have the requirement for training areas near the new force laydown locations.

Within the distributed laydown, the Marines must retain the ability rapidly to respond to crises across the range of demands, from major combat operations in northeast Asia to low-end humanitarian assistance and disaster relief wherever the crisis occurs.

Each location for the Marines is in transition as well. From Okinawa and Iwakuni, the Marines can locally train in Japan, Korea, and the Philippines, as well as respond with "fight tonight" capabilities if necessary.

From Guam, the Marines can train locally in the Commonwealth of the Northern Mariana Islands to the north, the Federated States of Micronesia to the south, and Palau and the Philippines to the west. Guam and the Commonwealth of the Northern Mariana Islands provide the Marines something we do not have anywhere else in the Pacific: a location on U.S. soil where they can train unilaterally or with partner nations.

In late 2011, President Obama visited Australia and launched with the Australians a new training relationship between the Aussie forces and the USMC. The Australian Prime Minister and the U.S. President highlighted the coming of the USMC to a training facility in the Northern Territory.

The visit provided a strategic opening for Darwin and the Northern Territory in the 21st-century approach of Australia and its allies to develop realistic training opportunities and thus establish war-deterring CONOPS.

Darwin's strategic location could become a hub of Pacific operations for Australia and a place to visit for its core allies. For the Marines, Darwin, Australia, allows them the opportunity to gain access to the large nearby training areas for portions of the year where we can conduct high-end, combined arms, and live-fire-and-maneuver training with a high-end ally.

By prepositioning appropriate equipment in Australia, the Marines could avoid the costly repetitive expense of moving equipment into and out of Australia while complying with Australia's bio-security measures.

And if another training facility located outside of Australia could be colocated with mobility assets, the Marines could then move people more easily to train in southeast Asia with partners. In other words, several moves are in play for the USMC in the Pacific.

The Marines are moving forces from Okinawa to Guam, building rotational forces to operate with the Australians in Australia, consolidating remaining forces in Okinawa, and moving some Marine forces forward from Hawaii into the western Pacific.

The danger would be that if these redeployments become aborted, the USMC position would be at risk. And funding needs to be available to ensure that the new force structure laydown is robust and effective.

Shaping the Transition: The Role of the Osprey

The Marines not only are physically moving in the context of the pivot to the Pacific but are introducing new equipment and capabilities in the region as well.

The Osprey has been introduced into Okinawa in spite of protests from residents of the island. The F-35Bs based at Yuma Marine Corps Air Station will deploy to the Pacific mid-decade.

A new flagship for the seabase, the USS America, will deploy to the Pacific in the next couple of years. The Marines are spearheading a relook at basing the region. With the evolving capabilities of the seabase — and the addition of new ships like the LPD-17s, the Joint High Speed Vessel, and the Mobile Landing Platform — and new concepts of the operation of the seabase, a foundation for shaping distributed capabilities is being laid.

And with it, the ability to mix and match land-based assets with sea-based assets will be important as well. The Ospreys can operate off of ships but operate with land-based assets to shape a joint intervention force. The F-35Bs can provide a capability to operate off of large-deck amphibious ships, large-deck carriers, and short runways on land.

And the F-35B carrying the same sensor and communications suites as USAF, USN, or allied F-35s can provide for a fleet engagement or leadership function.

The lead element in demonstrating the innovative possibilities of combining the new technologies with new concepts of operations has clearly been the Osprey.

As of September 2012, the Osprey reached a little noticed five-year mark in its operational deployment history. This aircraft, which can fly like a plane but land like a helicopter, has been a game changer for the USMC and its operations.....

Moving forward, we can see glimpses of the future that could lead to a cascading of change in operational approaches and capabilities if leadership will allow.

Three prospects for change are clearly evident already from the performance of the Osprey and its use in operations.

- First will be the impact of the "self-deployment" capability of the Osprey. The Osprey is able to with tanking fly directly to the area of operation. Try doing that with a helicopter. In fact, self-deployment is now being used in bringing Ospreys back from Afghanistan and used regularly in exercises. Self-deployment means that there is a possibility of rethinking how the seabase can work with land-based air. Ospreys can move with the fleet but be reinforced from land based Ospreys in plussing up air assault capabilities.

- Second is the impact of a new system like the Osprey on removing problems that threaten our warriors. There is a significant dimension to combat that can refer to problems avoided because of the performance and reliability of the new systems. The Osprey has avoided strikes that would have taken down CH-46s whether from manpads, RPGs, or other weapons fire. The Ospreys have proven robust in combat, where aircraft damaged by ground fire have used their digital management systems and redundant systems to self-correct and like the Timex watch in the ad, keep on ticking.

- The third will be the pairing of the F-35B, the first vertical-lift supersonic aircraft ever built with advanced sensors and C2 capabilities built in. The F-35B coupled with the Osprey will lead to a whole new level to begin shaping distributed operations over a large operational area.

SHAPING OPERATIONAL FLEXIBILITY: AN INTERVIEW WITH MAJOR GENERAL OWENS

January 20, 2013

The new technologies intersect with new approaches to create new options... In the case of the Osprey, the operation of the aircraft provides a very different way of thinking about basing options compared to older helicopters or rotorcraft.

The commanding officer of First Marine Air Wing, based in Japan, highlighted this change in an interview that he did with us. We discussed with Major General Owens recent exercises that his Marines conducted that presage changes in Pacific operations:

Question: Could you start by providing an update on Forager Fury?

MG Owens: *What was unique about Forager Fury was this was the first time we deployed MV22s in the exercise. This was just the first demonstration of the capability that aircraft brings to our AOR.*

And it went very well; the aircraft self-deployed, nonstop, supported by our KC130 aerial refuelers. They also worked in a fixed wing escort with the Hornets for a training opportunity we don't often get, complete with aggressor air en route.

Once we got the Ospreys to Guam, they did troop lifts, they did logistics flights; and then, the culmination was a Tactical Recovery of Aircraft and Personnel mission, to an island 200 miles away from Guam. We simply couldn't have done this with helicopters without doing front side and backside fueling stops in Tinian or Saipan.

With the Osprey, we were able to do it nonstop flying from Guam.

Question: There is a broader strategic point, which emerges from your exercise and the range and speed of the Osprey and the multiplier effect, which it and the coming F-35Bs could have on Pacific operations.

There are many islands in the Pacific. With the flexibility and relocation skills evident by the USMC (e.g., with regard to expeditionary airfields), islands can be a useful compliment to amphibious to provide the kind of presence which we may well need in the years ahead.

What is your thinking along these lines?

MG Owens: *This makes sense. We have a relative paucity of amphibious shipping. When I was a young lieutenant and captain, I think we had somewhere in the neighborhood of 65 amphibious war ships in the Navy inventory.*

Right now, we have 28 and they're spread about as thin as they possibly can be. We're running through their lifecycle faster than anticipated, and yet they're never enough. Going back to the whole challenge in this AOR is getting to where you need to be with some capability.

Being able to stretch the legs of the aircraft and operate from austere sites is critical.

A good case in point is that we just brought a couple of KC-130s back from disaster relief in the Philippines, a typhoon rolled through Mindanao and Palawan a few weeks ago. And we deployed a couple of KC-130s to haul relief supplies from Luzon to Mindanao.

The KC-130J was the aircraft of choice because there was a useable airfield at the southern end, at the affected end. But had there not been an airfield, which is often the case after tsunamis and typhoons, we could have done the same thing with the Osprey; flown it to Clark Field, operated out of Luzon — loading supplies in Luzon and dropping them to a point landing site in Mindanao supported by KC-130s in the air, providing aerial refueling.

And it's a capability we've never had before, and I expect that within the next couple of years, we'll have an opportunity to demonstrate that the Osprey may be the only aircraft that can get in to an affected area at the distance that we'll be required to support from.

Whether it would be from an intermediate staging base, like Clark or flying directly from MCAS Futenma here in Okinawa.

Comment: So in effect, an airborne infrastructure that allows you to have the reach and range to affect the situation on the ground.

MG Owens: *That is a good way to put it. When we put the KC-130 into the mix, we can bring some forward basing capability in the form of the maintenance crews that are required not only for the KC-130s, but also for MV-22s or whatever else that the tanker can drag to the objective area.*

If, in a time of conflict, we were going someplace and an adversary wanted to deny us the ability to put in a refueling point or intermediate support base, they would have to now take into account a much greater number of islands.

With only helicopters, an adversary could draw a 100-mile ring around a base and know where we could operate. Ospreys, particularly when supported by KC-130Js, would significantly complicate an adversary's attempts to predict our movements and operations.

THE OSPREY COMES TO THE PACIFIC: THE CASE OF THE 31ST MEU

May 10, 2013

The 31st MEU is the only permanently forward-deployed MEU and is deployed to the Pacific. It is also the unit that is working with 1st Marine Air Wing to generate Osprey deployments in the Pacific.

And given the centrality that Secretary of Defense Hagel has placed on Ospreys as part of the reinforcement of Japanese defense, the role of Ospreys and the deployment of 31st MEU in the region highlight some fundamental dynamics of change in the region.

This interview was conducted with Col Merna, the commanding officer of the 31st MEU in early May 2013:

Question: What is the plan for Ospreys to come to the MEU?

Col Merna: *VMM 265 will be chopped to us later this month. We are going to ease into the deployment much as was done with the East Coast MEUs to ensure that we execute wisely with the Ospreys. They will be part of our training with the Australians when we participate in Talisman Saber this summer.*

We will be training with them as well at Bradshaw Field, which is a training area, which is used during the rotational involvement of the Marines with the Australians. This training will contribute to the Australian effort to get ready to use their own forthcoming amphibious capability as well.

We are intending to operate with a full compliment of 10 Ospreys during the exercise, with 3 self-deploying from Okinawa, and we are steaming away with the rest of them. For us, the big deck amphibious ship will be the USS Bonhomme Richard.

Question: This is part of the process whereby the Osprey will become a normal part of Pacific defense?

Col Merna: *It is. There are clearly political sensitivities in the region, but the Japanese forces find the capability of interest and we are* working with the Japanese Ground Self Defense Force to familiarize them with the capability.

The potential sale of the Ospreys to the Israelis has made an impact in the region as well in terms of understanding the normalization of the Osprey as a key element of future defense capabilities. The options the aircraft provides us are significant.

For example, we can reach mainland Japan or the Philippines from Okinawa on one tank of gas, and, of course, with refueling the reach expands significantly.

This will also give our large deck amphibs a significant operational advantage as the Ospreys come onboard in the Pacific.

Comment: As the Japanese think through their evolving defense approach, they seem increasingly interested in the capabilities which the USMC and its blue team partners brings to the table.

Col Merna: *They are. During the last two cycles of our deployments, we have embarked Japanese Ground Self-Defense Forces with us to become more accustomed with our operations. We've integrate with them, we live with them, we train with them and certify with them.*

When we went to Thailand for an exercise, they came with us as well. They did remain aboard the ship during the exercise, it should be noted.

We are a maritime contingency force, which responds to any type of contingency ranging from humanitarian assistance to disaster relief to security operations and to higher end contingencies.

As such, we are key element in the Japanese perspective for their defense as well. Being out in the Pacific and engaging regularly and consistently, the 31st MEU, is extremely important to our Asia-Pacific strategy, not just combat ops on the Korean Peninsula.

It's an across the board presence and capability.

Question: In many ways, the Osprey is the most visible example of the transformation of Marine Corps operations. The F-35B will be very significant, but for the average Pacific citizen, they will see the impact from

the Osprey in very clear ways. Does that make sense to you?

Col Merna: *It does. When the next humanitarian assistance mission, disaster relief mission takes place in the Asia-Pacific, and we start impacting quickly, and immediately with relief supplies or people on the ground digging, filling sandbags, whatever it is helping somebody, all of our partners and allies in the Asia-Pacific region are going to see the immediate impact of the Osprey.*

And when they start seeing that big old ugly bird come flying in and dropping off supplies from areas that we could not reach previously, or as rapidly, they're going to see what they get out of the Osprey operating in their area.

Question: And for the full range of missions, the F-35B will be most visible in the lower end missions as its C2 and ISR mapping capability becomes evident in such crises.

That is why figuring out how to translate F-35B data to security operations will be a key requirement as well. But let us go back to the upcoming exercise with the Australians.

Could you discuss further?

Col Merna: *We're going to go down there and will do some live fire training for about a week. We are demonstrating to the Australians the impact of an amphibious capability.*

We're going to be able to take a battalion sized unit down the middle of nowhere, where there are saltwater crocs, there's nasty bugs, it's right in the middle of the outback that they very rarely use for military training.

And we'll use Ospreys probably in an insertion role, and set up everything in a C2 structure from the sea. That's the big piece here and if I can get the big deck to stick around, we're going to demonstrate really for the first time, an amphibious operational area southwest of Darwin.

We would be operating in a large area. The distance is around 400 miles and a minimum of 6 hours driving time. And then we're off the coast 70 miles at sea. And we will support a battalion size element training live fire for about a week, across all classes of supply, and then get them back on the ships.

And then, we'll take off from there. We will be exercising the range and scope of amphibious operations today. We're going to demonstrate an incredible capability across all classes of supply.

We will be able to be heavy on the aviation footprint, light on C2, but with significant operational capability to cover a significant area of operation.

Question: A final question would be how does the 31st MEU fit into the Pivot to the Pacific?

Col Merna: *In one sense, the Marines are going back to the force levels we had in the region prior to 9/11. So it is simply a restoration rather than a build up or buildout.*

But the way the force is being configured is very different. We are emphasizing building out a rotational force, notably in Australia, but elsewhere as well.

Because we are building out a rotational force, the new capabilities we are adding are crucial to success. Rotational forces require greater capability for reach and speed, key aspects of the Osprey-F-35B combination coming to the Pacific.

8

RANGE AND SPEED: THE SP-MAGTF

Next, there was the formation, deployment and then high demand use of what the Marines have called Special Purpose Crisis Response MAGTFs.

The first SP-MAGTF was formed in 2013 and leveraged the Osprey-KC130J combination to provide a force or what I would call later a "combat cluster" for supporting humanitarian or crisis interventions

In an interview done at the time of the initial standup of the capability in 2013 with Brigadier General James S. O'Meara then commander, U.S. Marine Forces Europe, and deputy commander, U.S. Marine Forces Africa, the role of the new force structure was explained.

In the companion volume to this book, I provide additional interviews from participants in the creation and operation of the SP-MAGTF as well. The reader will find there a number of insightful interviews there with regard to the operations of an SP-MAGTF which certainly highlights the unique contribution to what a V-22 enabled force can do.

According to O'Meara: *The SP-MAGTF is the basic Marine Corps air ground team or MAGTF approach but applied to a Special Purpose Mission.*

Special means it's uniquely tailored to a particular mission or a few mission sets.

In this case, the focus is upon security embassy reinforcements or a noncombatant evacuation.

Also, it is a rotational force, which provides a crisis response force able, to deal with EUCOM and AFRICOM needs.

General Dempsey provided strategic guidance, which was looking for a force, which operates with a small footprint, and is low-cost, and rotational. This is the answer to that guidance.

The SP-MAGTF meets the need to respond rapidly to a developing situation either proactively or reactively with a small force with a small footprint and has its own organic air, which means that it has operational reach as well.

The force is trained and operational and currently operating from a USAF base at Moran in Spain.

Question: The SP-MAGTF can reach into Africa or operate throughout the Mediterranean. Obviously, the Osprey is the enabler of such a force along with your organic lift and tanking. If you had only helos, this kind of force capability would not be possible, I would assume?

BGen O'Meara: *That clearly is correct. We can operate over a significant combat radius and of course, refueled with our C-130Js can reach throughout the region and all while carrying equipment, and/or two-dozen Marines inside. It gives AFRICOM commander a unique tailored operational tactical level force with significant operational reach.*

The V-22 allows for a paradigm shift and enables a force like SP-MAGTF. The V-22 gives you that C-130-like distance and speed with the versatility to land in confined, limited area, with no runway or an expeditionary LZ like a helicopter.

And when you add organic lift and tanking with our C-130Js, the reach is even greater and allows us to operate throughout Africa and the Mediterranean as needed.

And the self-deploying capabilities of the V-22 means that we can plus up the Osprey component as well as needed or other sites throughout the operating area.

Question: A key aspect of operating in Africa is of course shaping regional situational awareness and partnering skills. Could you talk to that important aspect of the force?

BGen O'Meara: *Indeed, this is an important part of the mission. With this template, we can send small training missions throughout the AOR with a small logistical footprint. The teams will visit for from 7-30 days to work with partner nations, as we are currently doing to prepare some African forces to work in Mali.*

There are actually three SP-MAGTFs in the region. For example, there is the Black Sea Rotational Force. Again, it's a MAGTF, with a battalion-sized company. And they're up in the Romanian area.

The Marines are clearly expanding the kind of partnership skills necessary for a highly mobile SP-MAGTF to be effective in coalition operations. We understand the region more than I think people assume we do. We understand the eastern part of Europe and now we're bringing the force south.

The flexibility of the force is important. It's an AFRICOM requested force in the case of the latest SP-MAGTF, but I think we have the flexibility, if we have to, to support EUCOM if needed it for whatever event pops up. It's, it's well-positioned forward.

And it provides a template, which is scalable.

It is a MAGTF and as such we can add key elements to bolster capability and over time the F-35B will be added to the mix, which certainly enhances the performance of this force template.

Comment: And you clearly can experiment with ways the SP-MAGTF can marry upon with the innovations the USN and USMC team are doing with the ARG-MEU as well.

BGen O'Meara: *That is a good point, but we are early in that process. What are the unique capabilities that we could tie together with that armed MEU, and aggregate them together to provide us even greater flexibility, and greater reach?*

Reach is a crucial aspect when operating in an AOR of the size we operate within.

Africa is a very large continent. You could take the Continental United States and put it inside the African continent. When I fly from Uganda to Stuttgart, it's eight-and-a-half hours via commercial air. That's the same as it is from here to Newark.

Also, the maritime regions of Africa are important in and of themselves and certainly for global trade and security.

The Operational Reach of Special Purpose MAGTF-CR

- With a KC-130J and Osprey Tandem, the USMC Special Purpose MAGTF Crisis Response Force Can Operate Over a Significant Operational Area.
- This is a significant change for ground forces from a rotorcraft-enabled and airlift force structure.

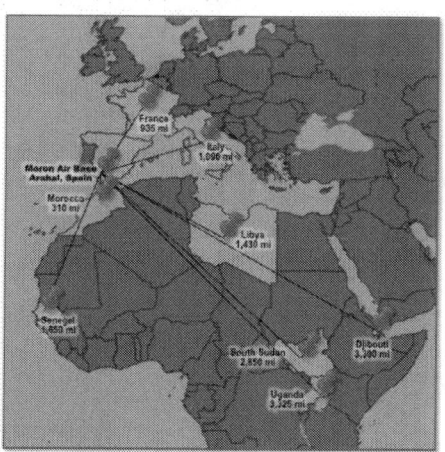

Credit: Second Line of Defense

❧ 9 ❧
THE HIGH-END FIGHT, CON-OPS AND FORCE DEVELOPMENT

In this chapter, I am including what I have characterized as two different but related phases of Osprey con-ops evolution. The first has been the shift from the land wars to the high-end fight which has required a significant shift in re-thinking and re-orientation.

And after that has come the focus on how to shape con-ops and forces which embody an ability to prevail in the high-end fight.

The first has been the strategic shift in orientation; the second has been working on practical ways to move in the new direction.

For economy of discussion, I will combine these two phases into one chapter.

The interviews in this chapter introduce the subject of the shift to the high end fight for the tiltrotor enterprise and the interviews and articles in the companion volume provide further details of the evolution.

THE STRATEGIC RESET

I have written several books since 2018 which deal with how significant has been the strategic shift from the land wars to dealing with multi-polar authoritarian powers and states, and notably with peer competitors.

I refer the reader particularly to my 2021 book. *Training for the High-End Fight: The Strategic Shift of the 2020s* for a discussion of the strategic shift.

But two interviews particularly highlighted how the shift was affecting the USMC and its thinking about its aviation assets. The first was an interview I did with the MAWTS-1 Commander in 2018, and the second was one I did the same year at 2nd Marine Air Wing with the Osprey squadron which participated in the Norwegian-led Trident Junction exercise.

In the interview with the CO of MAWTS-1 in 2018, I asked him the following:

Question: Clearly, there is a strategic shift underway for U.S. and allied forces to now operate in contested environments. That has happened during your time here.

How has that affected what you have had MAWTS-1 focus upon?

Col Wellons: *The team at 29 Palms as well as at Yuma have ramped up the contested and degraded environment that we present to our training audience at WTI and across all the other service level MAGTF training venues.*

We have challenged our students, especially this year, to operate in environments where communications and navigation systems are challenged, facing the most sophisticated and capable adversaries we can find.

We focused on the idea that in the future fight our primary means of navigation and communication will probably be denied, and certainly degraded and our operators may have to use old fashioned techniques to get bombs on target.

And the USMC ability to work force mobility was of growing

significance to enable the kind of distributed operations which was now being prioritized in the strategic shift.

Col Wellons commented: *Within the USMC, expeditionary operations are our bread and butter. In a contested environment, we will need to operate for hours at a base rather than weeks or months.*

At Weapons Tactics Instructor Training (WTI) we are working hard on mobile basing and, with the F-35, we are taking particular advantage of every opportunity to do distributed STOVL operations. It is a work in progress but at the heart of our DNA.

We will fly an Osprey or C-130 to a FOB, bring in the F-35s, refuel them and load them with weapons while the engines are still running, and then depart. In a very short period of time, we are taking off with a full load of fuel and weapons, and the Ospreys and/or C-130s follow close behind.

We are constantly working on solutions to speed up the process, like faster fuel-flow rates, and hasty maintenance in such situations.

Then in my discussions at 2nd Marine Wing about their participation in Trident Juncture 2018, a good sense of the strategic shift and what it meant for the Osprey was provided.

This is what I wrote in my article providing an overview on those discussions:

During my recent visits to Norway, I had a chance to visit several airbases and talk with a wide variety of Norwegian officers and defense officials.

With the return of direct defense challenges to the Nordics, there has been a major shift to recapitalizing the force, introducing mobilization measures and reworking the concepts of operations to deal with the Russian threat.

But it has been nearly two decades since the Nordics have faced a direct defense threat and at that time, they were facing the Soviet Bloc, and not simply Russia. This meant that the core threat they faced in times of war would be an amphibious assault from the Soviets similar to what the Germans did against them in World War II.

But now the threat is different and the concepts of operations not the same as well.

For the Nordics, the Trident Juncture 2018 exercise was a building block for shaping approaches to dealing with the new strategic situation.

The Norwegian Ministry of Defence described the exercise as follows:

"The exercise will test the whole military chain – from troop training at the tactical level, to command over large forces. It will train the troops of the NATO Response Force and forces from other allies and partners, ensuring they can work seamlessly together.

Why Norway?

"This exercise has air, sea and land elements, and Norway offers the possibility to train realistically in all of these domains. The cold and wet weather will pose additional challenges for NATO troops, and will train them to operate in extreme conditions.

"Norway offered to host Trident Juncture 18, and NATO accepted the offer more than four years ago. Norway has a long tradition of hosting major allied and multinational military exercises. Among them are Cold Response, Dynamic Mongoose and Arctic Challenge.

Why do we do exercises?

"Since 2014, collective defence has become a more prominent feature of NATO, due to the changes in the global security situation. In order to train and test NATO's ability to plan and conduct a major collective defence operation, the Alliance has held several large-scale exercises. This autumn, the turn has come to Norway.

"Trident Juncture is also a great platform to cooperate with close partners like Finland and Sweden – exchanging best practices and working together to address crises."[1]

During my visit to 2nd Marine Air Wing earlier this month, I had a chance to talk with Marines involved in the exercise to get their sense of the return of direct defense in Northern Europe and the challenges facing the Marines to provide the kind of force engagement which ultimately the Nordics, the U.S. and NATO would like to see in terms of coalition interoperability necessary to operate in a crisis situation.

MAG-26 and VMM-365 participated in the exercise. VMM-365 is an Osprey squadron.

I had a chance to talk with the CO of MAG-26, Col. Boniface, whom I have met with several times before he took this command, and LtCol Fowler, the CO of VMM-365.

According to Col Boniface: "It is important to note that during the exercise, which encompassed actions in Iceland and Norway, the V-22 operated above the Arctic Circle.

"We were able to deploy, engage and provide presence in the exercise. We had to deal with the weather and operating conditions in the region, which are quite different from where our Marines have spent most of their time in the past decade.

"And we need to continue to learn how to operate in those conditions, and to have the domain knowledge of how to exercise patience and timing appropriate to operations in the Nordic region.

"The weather comes in, each fjord has its own weather so to speak and we have to learn patience and how to deal with the second and third order affects which operating in cold weather generates."

Most of the conversation about the Trident Juncture 2018 engagement involving MAG-26 focused on the experiences of VMM-365 and LtCol Fowler provided an overview and various insights into the USMC experience.

According to LtCol Fowler, the impact of Hurricane Florence on North Carolina meant that they had reduced participation in the exercise. The initial plan was to send six aircraft, but they did send 114 Marines and 4 Ospreys to the exercise.

LtCol Fowler highlighted that they operated from the Iwo Jima amphibious ship and in Iceland did a raid against an "enemy" airfield. That raid was launched from the ship and the force returned to the ship after the raid.

The raid did not highlight the long-range capability of the Osprey but rather operated as integral part of the insertion force which also included CH-53Es and related assets.

A major piece of the operations in both Iceland and Norway was working with the Osprey in cold weather conditions. Notably, they were operating the Osprey's de-icing capabilities and getting a comfort level with the aircraft in cold weather conditions.

LtCol Fowler underscored the point made by Col Boniface with regard to the importance of weather conditioning and learning in Norway during the exercise.

"The Norwegians are great partners. They supported us as we worked our learning curve in the cold weather environment. But clearly we need to improve the communication systems used during the exercise, to get the full combat capability out of our force and to better integrate with the Norwegian force as well."

And as all pilots note when flying in Norway, it is not just the weather, which is challenging but the terrain and the infrastructure built into the terrain as well.

"With the towers and power lines running throughout the fjords, it is dangerous for aircraft operations. And we operate both as a helicopter and as an airplane so we faced challenges which are both the same but different for both type of craft all rolled up into one type of aircraft!"

"There was extensive use of UASs as well during the exercise, which creates a challenge to sort out the operations of the manned with the unmanned aircraft operating in the same airspace as well. Clearly, this is a work in progress."

One change which is critical to reshaping operations is the nature of the local community, meaning that when operating in Norway it was clear that they are a committed ally and the population was highly committed to supporting Marine Corps operations, including providing real time intelligence with regard to the "enemy" force. This was noted as a significant difference from USMC operations in the Middle East.

In short, the picture provided of MAG-26 involvement in Trident Juncture 2018 reinforced the picture provided by MAG-31. The exercise was a success in terms of being able to project force, but to get the full combat value from a Marine Corps force in a real crisis, significant effort needs to be directed towards enhanced capabilities to integrate the insertion force with the host nation and its force.

DISTRIBUTED MARITIME OPERATIONS AND FORCE DEVELOPMENT

A major focus within the strategic shift has been upon the joint and allied forces working various forms of distributed operations. The Navy focus has been upon distributed maritime operations; the Air Force it has been upon Agile Combat Employment, and for the USMC it has been upon distributed operations integrated with the joint force, notably the Navy.

The Osprey has been a key element of working this force re-design effort. It has come to the U.S. Navy as a key element for providing for fleet support and intra-theater logistics; it has expanded its multi-mission capabilities as the notion of force distribution supported by air enabled capabilities with a variety of payloads has been evolving.

Let me first address the coming of the Osprey to the fleet to provide for its supply function and then address evolving thinking about multi-mission payload enablement.

In other words, there are two ways to look at the evolution of the tiltrotor enterprise addressing the high-end fight — how the Navy has adopted it for the significant contested logistics role and how the Marines are working the multi-mission every-thing approach to using the Osprey as a very capable combat platform able to take the payloads of the evolving kill web force.

THE IMPACT OF THE U.S. NAVY JOINING THE TILTROTOR ENTERPRISE

The importance of the Osprey in the shift in con-ops of the U.S. Navy was highlighted in an interview I did in April 2023 with a senior U.S. Navy commander. He emphasized that the Navy needs to be able to more effectively do intra-theater logistics and to do so with effective speed and survivability.

As the Admiral underscored: "If we are going to have distributed maritime operations, we better have the ability to support, battle damage repair, sustainment, and medical services provided at way more rapid than 20 knots."

With the Navy needed to augment significantly over time its intra-theater logistics support, they are starting with the replacement of the C-2A Greyhound with the CMV-22B tiltrotor aircraft. The Admiral described this as a shift from a limited specialized support asset to having a distributed fleet support asset which provides for intra-theater logistics, a priority need.

Currently, the Navy plans to only replace C-2As with Osprey numbers pitched to carrier peacetime deployments. But not only is the question of needs for logistics in contested operations, but also broader intra-theater logistics raises questions about the real numbers required. As he noted: "I think also we might be more creative in our Osprey Con Ops, which would enable us to unlock some of our thinking on what else can be done."

He conceptualized the shift as follows: "The COD was a tactical capability. The COD had a built-in operational inefficiency. Logistics was executed at the operational level. This unit needs X, Y, and Z based on their consumption. We then estimated they're going to need X, Y, Z, and planned to these levels. We then assigned the COD as a small unit with low density and high demand supporting only carrier strike group, so that we could fill in the gaps that operational planning doesn't account for and it will be a uniquely oriented machine that can land on our aircraft carriers and it can land on land.

"With an Osprey, we have a bigger footprint so that we are no longer a low density, high demand asset. Because the COD was a small, exquisite capability that we're only going to use for one type of platform, we had small numbers which in turn affected availability of the asset. With the Osprey, we have all manner of platforms we can land on or service stations we can operate from. We are not committed to a tail hook and its limitations. This gives us logistics capability which allows us to make the choices about who needs to be resupplied and at what time, day, or night."

The Osprey provides an important stimulant for the shift in con-ops whereby the Navy's experimentation with distributed operations intersects with the U.S. Air Force's approach to agile combat employment and the Marine Corps' renewed interest in Expeditionary Advanced Base Operations (EABO).

In other words, the reshaping of joint and coalition maritime combat operations is underway which focuses upon distributed task forces capable of delivering enhanced lethality and surviv-ability.

The U. S. Navy's deployed fleet — seen as the mobile sea bases they are — faces a significantly different future as part of a distributed joint force capable of shaping a congruent strike capability for enhanced lethality. This means not only does the fleet need to operate differently in terms of its own distributed operations, but also as part of modular task forces that include air and ground elements in providing for the offensive-defensive enterprise which can hold adversaries at risk and prevail in conflict.

The Navy's version of the Osprey — the CMV-22B — is ideally suited to operate across this highly complex distributed combat chessboard. And, because the Marines have deployed the MV-22B for decades, there is a very robust operational and sustainment expertise already in the fleet. This means the CMV-22B can deliver core carrier logistics needs while also providing logistics support across the entire fleet — including the vital Military Sealift Command that will play an essential role.

As the fleet looks to enhance its lethality and survivability in a distributed maritime environment, there is no more critical capability than sustained logistics support in the contested battlespace.

This is how Rear Admiral Meyer, Commander, Naval Air Force Atlantic, put it in an interview with me in regard to how the Navy was reworking carrier operations in a way that high-lighted this key logistics requirement: *The fact that our carrier strike groups can move 700-plus miles in a 24-hour period, the increasing range and lethality of our ever-advancing air wing and the weapons that those aircraft carry can hold huge areas of the surface at risk.*

Over the course of a three-day period, it would mean just a staggering volume of real estate, roughly the entire Pacific AOR over a 72-hour period. But it is that logistics support train that is really a key part that makes that happen.

The CMV-22B can do this for the carrier-enabled distributed maritime force.

A MISSION-KITABLE AIRCRAFT FOR KILL WEB OPERATIONS: COLONEL MARVEL DISCUSSES THE WAY AHEAD WITH THE OSPREY

February 24, 2023

In my book with Ed Timperlake on the coming of the maritime kill web, we underscored that a distributed force built around modular task forces highlights the payloads which the task force can deliver in terms of effects, rather than describing that task force in terms of a core platform.

The flexibility which the Osprey provides – with the USMC, the U.S. Navy and the USAF operating the aircraft – opens the aperture significantly on how one configures the aircraft to deliver what payload in which situation for which combat and deterrent effect.

To understand more about this change for the Osprey Nation, I recently talked with Colonel Marvel, the CO of MAG-39, located at Camp Pendleton. Col Marvel and his team have worked closely with the CMV-22B team at North Island in the standup of the Navy's Osprey capability.

Col Marvel underscored that expanding the mission set for the Navy's CMV-22B was certainly possible but was not in his domain of responsibility. But the USMC is clearly expanding the payloads carried by the MV-22B which supports distributed operations, and if the three services which operate the aircraft found ways to expand their ability to cross-service each other's aircraft, they would be able to enhance such operations.

As Col Marvel put it: *The Osprey provides unique speed and range combinations with an aircraft which can land vertically. It is a very flexible aircraft which could be described as a mission-kitable aircraft. The Osprey has big hollow space in the rear of the aircraft that can hold a variety of mission kits dependent on the mission which you want the aircraft to support.*

A U.S. Marine Corps MV-22B Osprey with Marine Medium Tilt Rotor Squadron 165, Marine Air Group 16, 3rd Marine Aircraft Wing (MAW), refuels at a forward arming and refueling point Dec. 1, 2022, on Camp Wilson, Marine Corps Air Ground Combat Center Twentynine Palms, Cailfornia, during Steel Knight 23. Exercise Steel Knight 23 provides 3rd MAW an opportunity to refine Wing-level warfighting in support of I Marine Expeditionary Force and fleet maneuver. (U.S. Marine Corps photo by Lance Cpl. Jacob Hutchinson)

He emphasized that with a variety of roll-on roll-off capabilities with different payloads: *We can add the specialists in the use of a particular payload along with the payload itself to operate that payload, whether kinetic or non-kinetic, whether it is a passive or active sensor payload. We need to stop thinking about having to put the command of such payloads under the glass in the cockpit, and control those payloads with a tablet.*

Col Marvel underscored that the Marines when deployed are engaged in presence missions. How then best to use their presence to deliver the desired effect? And given the Marines are operating across the spectrum of warfare, and that spectrum itself is changing, which payloads are most relevant to the mission? According to Col Marvel: *This means that we need to maximize the payload utility of our platforms.*

He provided a number of examples of different payloads which they are working with from USVs to a variety of passive and active sensors. Kill webs need to be sustained and Ospreys can provide both fuel and ordinance to platforms throughout the extended battlespace. For example, Ospreys can bring fuel and ordinance to a FARP (forward arming and refueling point) and support P-8 operations, for example.

Ospreys can palatize torpedoes and engage them in the battlespace. They can provide key parts of the network of sensors that make a distributed forces' domain C2 and fires control picture. With the proper payload, Ospreys can maintain contact with surface and subsurface forces to help build a common tactical operating picture.

The Navy with the USMC are doing a wide range of mission rehearsal experimentation to determine how best to operate a variety of payloads operating off of various platforms to enable the distributed force to have the kind of effect – kinetic or non-kinetic – which is desired. The Osprey – with its range, speed, and landing flexibility – can play a key role in this overall effort.

One example of the kind of experimentation which Col Marvel was talking about was highlighted in a 3 January 2023 story released by 3rd Marine Aircraft Wing.

This story highlighted actions by Marine Air Control Group 38 in Exercise Steel Knight 2023.

Marines and Sailors with Marine Air Control Group 38 refined tactics for future maritime conflicts in the Indo-Pacific region during Exercise Steel Knight 2023. With units positioned across California and Arizona, MACG-38 tested components of Aviation Command and Control in conjunction with 3rd Marine Aircraft Wing's "Hub, Spoke and Node" model in preparation for the next fight.

Acting as the "Hub," MACG-38 established a Tactical Air Command Center aboard Marine Corps Base Camp Pendleton, San Diego, California. With the TACC fully operational, MACG-38, in conjunction with 3rd MAW key leaders, facilitated command and control of air assets throughout the battlespace.

Concurrently, MACG-38 set up a "Spoke" at the Strategic Expeditionary Landing Field at the Marine Corps Air Ground Combat Center located in 29 Palms, Ca in order to facilitate Marine Aircraft Group 16's Assault Support requirements for Steel Knight 23. From the SELF, Marines from MACG-38 were able to establish

Forward Arming and Refueling Points were supported by MV-22B Ospreys and CH-53E Super Stallions from MAG-16. From these remote locations, MACG-38 Marines were able to support the refueling of both MAG-16 aircraft and tactical air platforms including the F-35B Lightning II from MAG-13.

Off the coast of San Diego on San Clemente Island, MACG-38 also embedded a platoon of Marines from 3d Low Altitude Air Defense Battalion with the 11th Marine Regiment in order to earn certification for the upcoming Marine Rotational Force-Darwin deployment to Australia.

While on San Clemente Island, LAAD Marines provided critical support for a large-scale Air Assault. During the exercise, LAAD gunners executed over a dozen live-fire shoots and earned their certification for MRF-D.

Steel Knight 23 also saw the establishment of organic Air Control Companies within 3rd MAW. In order to experiment with and force generate Multifunction Air Operations Centers without impact to global force management tasking, MACG-38 transferred personnel and equipment from MACS-1 Air Defense Company Bravo to MASS-3 in order to reorganize Marine Air Support Squadron 3 into three Air Control Companies and a Headquarters Company.

Under this new construct, MASS-3 was both able to support all I Marine Expeditionary Force and 3rd MAW C2 functionalities and requirements over a broader spectrum by capturing MAOC personnel and equipment requirements, techniques, tactics, and procedures, and training requirements.

This initiative led to the Initial Operational Capability of MASS-3 Air Control Companies as well as the MAOC force generation event supporting 1st Marine Regiment's MRF-D deployment.[2]

In short, as Col Marvel underscored: *There's a lot of capability that we have now. And our fight is today; today is our last day of peace out in the fleet.*

❧ 10 ☙

AN OVERVIEW ON THE
COMING OF THE CMV-22B TO
THE LARGE DECK CARRIER

I have had several interviews with U.S. Navy officers involved with the CMV-22B replacing the C-2A in the carrier resupply role. An article which I published on October 1, 2020 provided a useful summary on the coming of the CMV-22 to the fleet as seen through interviews which I had conducted previously.

In the companion volume to this book, I provide a number of interviews which I have conducted through the end of 2024 on the importance of the new Osprey for the Navy's contested logistics mission across the fleet.

This chapter was published 1 October 2020 by me and provides an overview as of that time of the coming of the CMV-22B.

That article follows:

October 1, 2020

I first viewed an CMV-22B in person when attending the Reveal ceremony in Amarillo, Texas held on February 6, 2020.

But I am not stranger to the Osprey having seen my first Ospreys at Second Marine Air Wing in 2007.

At that time there were four Ospreys on the tarmac.

The plane has come a long way since then with the Marines

taking it in to every clime and place with the transformational aircraft having a significant change on how the Marines operate.

Now the plane is coming in a modified form to the US Navy, and it is technically replacing the C-2 Greyhound in its carrier support role. The CMV-22B is no more a replacement for the C-2 Greyhound, than the MV-22 was for the CH-46. The MV-22 covered the functions of the CH-46 for the Marine Corps but represented a disruptive change which has transformed the USMC and its operations.

The CMV-22B will provide the functionality of the C-2 for the carrier strike group but is entering the carrier strike group at a time of profound change, and it will contribute to it.

Over the past few months, I have had a chance to discuss the coming of the CMV-22B to the large deck carrier with a number of people knowledgeable about the transition and would like to share those findings.

I have travelled to Pax River, San Diego, Naval Air Station Fallon, and Amarillo, Texas over the past few months, and would like to share what I have learned from those visits for these findings form a baseline with regard to the importance of the coming of the CM-22B to the fleet.

PAX RIVER

I went last Fall to Naval Air Station Patuxent River and to meet with Col Matthew Kelly, who is in charge of the V-22 Joint Program Office (PMA-275). I first met "Squirt" when he was an F-35B test pilot and indeed was selected as a test pilot of the year in 2011. Having come from the F-35 world, where the entire command and control (C2) and intelligence, surveillance and reconnaissance (ISR) infrastructure is being reworked, is a perfect community for the new head of the V-22 Program to come as that aircraft is undergoing a fundamental transformation.

It is often overlooked that the U.S. Air Force Special Forces

Command (AFSOC) and Marine Corps are still the only tiltrotor forces in the world. And the Osprey from the outset has demonstrated a speed and reach capability which traditional rotorcraft simply have not replicated.

At Pax River, we discussed the next phase of the evolution of the Osprey of which now the U.S. Navy's carrier community would become a key player as well as the Marines and the US Air Force.

The U.S. Navy is joining Osprey Nation at the same time as the Japanese. As Col. Kelly commented: "There is no other air platform that has the breadth of aircraft laydown across the world than does the V-22. And now that breadth is expanding with the inclusion of the carrier fleet and the Japanese.

"We currently have a sustainment system which works but we need to make it better in terms of supporting global operations. With the U.S. Navy onboard to operate the Osprey as well, we will see greater momentum to improve the supply chain."

DISCUSSIONS WITH THE U.S. NAVY'S AIR BOSS

After Pax River, my next discussion of the coming of the CV-22B was with the Navy's Air Boss, Vice Admiral Miller. In a meeting in his office in San Diego a week before attending the reveal ceremony in Amarillo Texas, we discussed how the Air Wing was changing, and the impact of the coming of new platforms, like the CMV-22B. As we discussed the future of the air wing, we agreed that a way to look at the way ahead was not so much the integrated air wing, but the shaping of the integratable air wing.

What is being set in motion is a new approach where each new platform which comes into the force might be considered at the center of a cluster of changes.

The change is not just about integrating a new platform in the flight ops of the carrier. The change is also about how the new platform affects what one can do with adjacent assets in the

CSG or how to integrate with adjacent U.S. or allied combat platforms, forces, and capabilities.

We then focused on the case of the U.S. Navy replacing the C-2 with the CMV-22 in the resupply role. But the Navy would be foolish to simply think in terms of strictly C-2 replacement lines and missions. So how should the Navy operate, modernize, and leverage its Ospreys?

For Miller, the initial task is to get the Osprey onboard the carrier and integrated with Carrier Air Wing (CVW) operations.

But while doing so, it is important to focus on how the Osprey working within the CVW can provide a more integrated force.

Vice Admiral Miller and his team are looking for the first five-year period in operating the CMV-22 for the Navy to think through the role of the Osprey as a transformative force, rather than simply being a new asset onboard a carrier.

THE REVEAL CEREMONY

After my visit with the Air Boss, the following week I travelled to Amarillo Texas for the reveal ceremony. My assessment at the time was that I was the only outsider in attendance and had a chance while at the ceremony to talk with a wide range of attendees from the USMC, the U.S. Navy, industry and the acquisition community. But I will underscore three interactions which highlight the way ahead for the U.S. Navy and its CMV-22B.

The CMV-22B unveiled at the ceremony,

The first interaction was with Capt. Dewon "Chainsaw" Chaney, the Commander of COMVRMWING (or Fleet Logistics Multi-Mission Wing) who command three squadrons of CMV-22Bs as they are stood up. At the ceremony, Captain Chaney highlighted the coming of the new capability and what it meant for the US Navy.

"What is the status of the CODs? Every Carrier Air Wing Commander and Carrier CO has received that question numerous times from the Carrier Strike Group Commander while on deployment. And for good reason...

"The COD, or Carrier On-board Delivery, aircraft is the only long-range aerial logistics platform providing logistical support for the Carrier Strike Group, ensuring its time sensitive combat capability. Sure, there are ways to get some items to the carrier but that time lag in most cases is at the cost of readiness for the warfare commander.

"The Navy saw the need to replace the aircraft providing this critical capability years ago and embarked on multiple efforts to inform that decision. The Navy selected V-22 as the future COD platform. The first aircraft is being delivered today (well actually a week ago but who is counting). And our first deployment will be here in a blink of an eye!

"But the devil in the details with this particularly accelerated program is making sure that the fleet can man, train, and equip those at the tip of the spear potentially in harm's way.

"As of October, last year as the Wing Commodore, I have the honor, privilege and responsibility, given to me by Vice Admiral Miller, to be the lead for the Navy's CMV-22 community along with our partners at well into the 2040s. Delivery of this aircraft is a major milestone on the path to initial operational capability in 2021.

"The CMV-22 has the capability to internally carry the F-35C engine power module. This capability is a game changer for the Air Wing of the Future and drove the need to match up the F-35C and CMV-22 operational deployments. The first CMV-22 deployment is now less than a year from initial delivery of N3, which is scheduled for late June of this year.

"Its success is key to maintaining combat lethality for the Air Wing of the future and our Navy. CMV-22s will operate from all aircraft carriers providing a significant range increase for operations from the Sea Bases enabling Combatant Commanders to exercise increased flexibility and options for warfare dominance.

"If you're in a fight, it's always good to have options! Every month following the first initial deployment, there will be a CMV-22 detachment operating with a U.S. aircraft carrier somewhere in the world...."

THE PERSPECTIVE OF CAPT (RET.) SEAN MCDERMOTT

The second interaction was with CAPT (ret.) Sean McDermott who currently is a commercial airline pilot who served in the U.S. Navy for 26 years. He was involved with the C-2 during the majority of his career, starting as a Greyhound pilot and eventually commanding one of the Navy's two fleet logistics squadrons. In the final years of his service, McDermott was involved in working through options for the Navy as they considered C-2 replacements, with an eventual Osprey selection.

McDermott highlighted the potential for the CMV-22B to expand the envelope significantly for what a COD aircraft could do for the fleet.

"With the C-2 we did one thing – Carrier On-board Delivery. With the Osprey, Combatant Commanders already know the multi-mission capability of the V-22 and will be tempted to utilize them for a variety of other missions.

"This is not something that would happen with a C-2. Carrier leadership will eventually struggle to fence off their logistics assets from outside tasking."

In other words, there is an anticipated operational demand

that they will want to leverage fully the new versatile capabilities of the Osprey.

He noted that with the new platform being introduced to carrier aviation, it will be possible to leverage it to shape a greater range of capabilities for the COD asset. He noted that as the Marines began to get comfortable with the MV-22, they shaped the unique Special Purpose Marine Air-Ground Task Force (SP-MAGTF), which has become a highly demanded asset.

He argued that such innovation was certainly possible for the Navy as it worked with its new COD aircraft.

One area he noted were forward deployed locations that would benefit like operations in Bahrain. Ospreys deployed to these locations could not only better support logistics but would also have the flexibility to support other mission sets for combatant commanders.

"With the coming of the new platform into the fleet, one innovation which might be considered is how to use the new Navy Osprey as part of a broader sustainment effort encompassing Marine Corps and Navy Ospreys. It also is an area where the multi-mission capabilities of the aircraft for the Navy can be explored as well.

"In other words, where the Marines leveraged their Ospreys to build and equip SP-MAGTF, perhaps the U.S. Navy can leverage the Bahrain anchor from which to build regional sustainment and explore ways to build out the multi-mission capabilities it would want from its CMV-22s."

This clearly might require the Navy to consider from the outset ways to ramp up the buy and to prepare for ways in which the fleet commanders will employ it to leverage fully the aircraft capabilities, and, at the very least, utilizing its capability to provide improved logistics to Navy and Maritime Sealift Command ships.

THE PERSPECTIVE OF THE MAYOR OF AMARILLO

The third interaction of note I had in Amarillo was with its Mayor. Neglected in any discussion of new capabilities is the contribution of the workforce which builds such a capable aircraft as the Osprey. And I asked the Mayor, why is Amarillo, Texas capable of doing so.

To be blunt, I asked here the following question: From where are these skilled workers coming from, and why is Bell here?

Mayor Nelson is the third from the right in a photo after the reveal ceremony.

Mayor of Amarillo, Ginger Nelson, provided a spirited response:

"Because we want Bell here, because we have a tremendous workforce here in the Texas Panhandle," Nelson said.

"We are a city fed by the small-town rural communities that surround our region. Our people are only one or two generations from having grown up on a farm or having owned their own small business. And the work ethic for our people is simply: if you are not doing it, it is not going to get done. Our work ethic is strong; and patriotism is a core value in the Texas Panhandle."

The often-forgotten enabler of the U.S. military is the industrial worker. But Ginger Nelson certainly has not forgotten their importance. "Bell relies on us to supply dedicated, competent workers who are ready to meet the responsibilities that include the defense of our nation."

VISITING NAWDC

After my visit to Amarillo, my next visit involving discussing the way ahead with regard to the CMV-22B was at Fallon Naval Air Station, the home of the Naval Aviation Warfighting Development Center (NAWDC). This key training center is hooked up with other Navy and warfighting centers to generate the kind of innovative combat force which can defend the nation's interests.

And new Navy air platforms coming into the force are vetted into NAWDC to shape their maximal contribution to a lethal and effective combat force. The C-2 never was a plankholder in NAWDC, but the CMV-22B will be.

According to CO of NAWDC< Rear Admiral Brophy, they will work the tactics, techniques, procedures (TTPs) for the CMV-22Bs along with Captain Chaney, as it will enter into NAWDC through the rotary wing school in NAWDC, but its ultimate location for cross-platform training, in a command increasingly focused on such training with a kill web focus, will be determined.

VISITING THE NORTH ISLAND AIR STATION, SAN DIEGO

After my visit to NAWDC in early July 2020, I went to San Diego and met with the Naval Air Boss on the morning of July 13th and in the afternoon with "Chainsaw." During my visit with the Osprey squadron, I had a chance to see the third Osprey on the tarmac, and visit the hangar being used to stand up the squadron.

In my discussion with "Chainsaw" at North Island on the 13th of July 2020, we discussed the standup of the CMV-22B squadrons.

The first squadron VRM-30 was stood up prior to the creation of the wing and its first aircraft arrived in June 2020. Captain Chaney noted that there is a two-year timeline to get a fully qualified maintenance technician or officer for the force, so that has been underway.

The photo is of Captain Dewon "Chainsaw" Chaney and myself during a visit to North Island in 2021.

That training has been generated with the Marines in North Carolina, Hawaii, Kuwait, or working side by side with Marines in various locations or in the Bell-Boeing teams at the Maintenance Readiness Team in Miramar.

Captain Chaney then noted that this October, the fleet replacement squadron, VRM-50, will be stood up. It will take this squadron two years until they will be able to train new pilots. As he explained: "With VRM 30, they need to get pilots ready to go fly and go on deployment."

"Whereas with VRM 50, they have to get pilots and maintainers qualified, but then they also have to figure out how to train other pilots and aircrewman in other words to establish the Navy training cycle for the aircraft."

He noted that the Navy will approach operating its Osprey in some ways differently from the Marines, but because of the interactive working relationships any learning on the Navy side can be easily be transferred on the Marine side. "I see it as a very symbiotic relationship between the Marine Corps and us, all under the Department of the Navy.

"Clearly with the Marine Corps having the bulk of the experience right now in MV-22s, I welcome any of their lessons learned and comments about maintaining the airplane, flying the airplane, fighting with the airplane. I'm all ears, because I know that my team is still in their infancy."

But one example of cross learning might be with regard to how the Navy will operate the load outs and off-loading of the aircraft. They are looking to have a rapid unload capability with new containers for the CMV-22B and Navy experience with the new kit might well prove of interest to the Marine Corps as well. The counterpart to VRM-30 will be VRM-40 but all three squadrons are under the Wing. The third squadron will be on the East Coast.

And as the Osprey comes to the fleet, building appropriate infrastructure is a key priority facing the Wing in the next few years. At North Island, San Diego, their first simulator will come next year, and a new hangar is being built and will be ready in 2023. But the East Coast basing solution remains to be resolved.

With regard to the standup, the Wing Commander comes from the rotary wing community; his Deputy from the C-2 community.

The challenge is blending the two into a tiltrotor force which operates at a different altitude from the C-2, can fly night shipboard missions (which the C-2 did not), and rapid, efficient shipboard operations, which has not been the core focus of the USMC and their use of the aircraft.

In short, it is clear that the CMV-22B needs to prepare for carrier operations but equally the carrier community needs to get ready for the coming of CMV-22B.

AN OCTOBER 2024 UPDATE ON
THE EAST COAST CMV-22B
SQUADRON

I n this chapter, I am going to address, a visit to Norfolk in the Fall of 2024 which highlights the coming of the Osprey to provide core functions for the United States Navy. I visited Fleet Logistics Multi-Mission Squadron (VRM) 40 the "Mighty Bison" on 29 October 2024.

As I walked to the squadron's temporary facility and looked at the location of where their new hangar is being built, I thought back to the last time I encountered an Osprey at Naval Station Norfolk.

And that is when I landed there on a USMC Osprey when returning from HMS Illustrious at sea off the Virginia coast.

That was 17 years ago, and much has transpired in the tiltrotor enterprise since that time which is the focus of this book.

17 years ago, we took off from the Pentagon to fly to the HMS Illustrious but when we were to return the weather was threatening and we landed just after a Hawkeye did at the airfield at Naval Station Norfolk.

This was I wrote earlier about that experience in *Military Logistics International* story published in September-October 2007:

In July of this year, the USMC assigned two Ospreys and fourteen AV-8B Harriers to operate aboard HMS Illustrious. The British aircraft carrier was participating in a joint exercise with the U.S. and other allied navies near the Virginia and North Carolina coasts. The exercise was an unprecedented effort by the Marines and the Royal Navy, in which close coordination allowed the Marines to operate fully off the British ship....

For the USMC, the exercise provided an opportunity both to certify pilots and, more importantly, to develop coalition operational skill sets. The USMC is a flexible fighting force and sees its range of missions as requiring the ability to work with allies at sea and on land. The preparation for the exercise and the experiences of the exercise itself allowed the Marines to work closely with the Royal Navy.

And to thereby further develop coalition collaborative combat skills. It was not a technical exercise in interoperability: rather the Marines saw the exercise as an opportunity to develop an on-the-fly-division of labor skill sets so necessary for coalition operations. British procedures were mixed with Marine Corps procedures in crafting a blended coalition combat capability

Of course, the Brits and the U.S. Navy and Marines have gone on since that time to integrate the new UK carriers with U.S. fleet operations. And the Osprey has been a lynchpin to getting that process further advanced.

In the photo below, the Osprey we took to the ship and which we took from the Pentagon helo pad can be seen.

And in the next photo, the second Osprey which flew with us can be seen landing on HMS Illustrious.

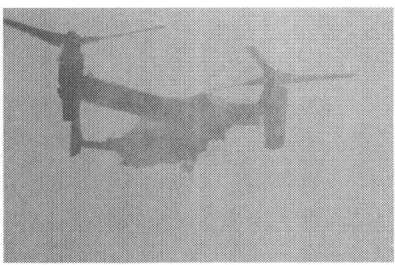

But now Naval Station Norfolk has its own Osprey squadron, but one not yet envisaged 17 years ago, namely a logistics enabler for the fleet. This Osprey has been redesigned to add significant additional fuel capability for its fleet support mission, centered around support for the aircraft carrier.

Squadron patch and challenge coin given to me when I visited the squadron.

This is how its arrival was highlighted in a press release by the Commander, Naval Air Force Atlantic Public Affairs published on 3 May 2024:

Fleet Logistics Multi-Mission Squadron (VRM) 40 the "Mighty

Bison" held a change of command ceremony aboard Naval Station Norfolk, May 2. Cmdr. Matthew Boyce, who is from Spokane, Washington was relieved by Cmdr. Mason Fox, who is from Lemoore, California served as the executive officer before assuming the position as commanding officer during the ceremony.

The first East Coast-assigned Navy tiltrotor vertical/short takeoff and landing (V/STOL) CMV-22B Osprey aircraft, assigned to VRM-40, arrived in Norfolk on April 5. The remaining VRM-40 aircraft will begin to arrive to Hampton Roads in the weeks to come...

The CMV-22B will provide the fleet's medium-lift and long-range aerial logistics capability, eventually replacing the C-2A Greyhounds of Fleet Logistics Support Squadron (VRC) 40 over the next several years. The squadron's relocation to Naval Station Norfolk is part of their permanent duty station change from Naval Air Station (NAS) North Island in preparation to provide fleet logistic aviation assets to the Atlantic Fleet beginning in 2025....

Naval Air Force Atlantic is responsible for seven nuclear-powered aircraft carriers, 55 aircraft squadrons, 1,200 aircraft and 52,000 officers, enlisted and civilian personnel with priorities focused on warfighting, people, and readiness by providing combat ready, sustainable naval air forces with the right personnel, properly trained and equipped, with a focus on readiness, operational excellence, interoperability, safety, and efficient resourcing.[1]

THE ARRIVAL OF THE CMV-22B TO NAVAL STATION NORFOLK

In an April 5, 2024 story released by the Commander, Naval Air Force Atlantic, the arrival of the first CMV-22Bs to Norfolk Naval Air station was highlighted.

The first East Coast-assigned Navy tiltrotor vertical/short takeoff and landing (V/STOL) CMV-22B Osprey aircraft, assigned to Fleet Logistics Multi-Mission Squadron (VRM) 40, arrived to Naval Station Norfolk on April 5.

"Naval Aviation is ecstatic to welcome the first CMV-22B Osprey to

Norfolk," said Rear Adm. Doug Verissimo, commander, Naval Air Force Atlantic (CNAL). "This first aircraft's arrival symbolizes an evolution and change in Naval Aviation as we look toward the future. The event represents the hard work and stamina of our aviators, aircrewmen, maintainers and sustainment personnel in the VRM community."

The first East Coast-assigned Navy tiltrotor vertical/short takeoff and landing (V/STOL) aircraft CMV-22B Osprey conducts post-flight checks following its arrival at Naval Station Norfolk, April 5, 2024. U.S. Navy photo by Mass Communication Specialist Seaman Sylvie Carafiol

The CMV-22B will provide the fleet's medium-lift and long-range aerial logistics capability, eventually replacing the C-2A Greyhounds of Fleet Logistics Support Squadron (VRC) 40 over the next several years. The squadron's relocation to Naval Station Norfolk is part of their permanent duty station change from Naval Air Station (NAS) North Island in preparation to provide fleet logistic aviation assets to the Atlantic Fleet beginning in 2025.

The VRM-40 "Mighty Bison" were established aside their existing sister squadron, VRM-30, and the training squadron, VRM-50, aboard NAS North Island in March 2022.

All squadron personnel have been officially stationed in Norfolk since Feb. 1, 2024. The remaining VRM-40 aircraft will begin to arrive to Hampton Roads in the summer of 2024.

VRM-40's leadership consists of Cmdr. Matthew Boyce,

commanding officer; Cmdr. Mason Fox, executive officer, and Command Master Chief Bradley Wissinger.

"We are proud to join the Commander, Naval Air Force Atlantic team and eager to lean forward into our next phase of stand-up," Boyce said.

Fox discussed the importance of standing up a new squadron on the East Coast.

"We're excited to be in our permanent home at Naval Station Norfolk and focused on continuing to build the squadron to execute our mission — delivering high priority people and parts to carrier strike groups at sea," Fox said. "The Osprey is an extremely capable aircraft and will be critically important to the way the Navy fights for many years to come."

In addition to VRM-40, a type wing detachment was established onboard Naval Station Norfolk earlier in 2023 to provide local representation of Commander, Fleet Logistics Multi-Mission Wing (CVRMW), based at NAS North Island.

CVRMW's mission is to provide Pacific and Atlantic Fleet VRM squadrons the ability to sustain lethality for carrier strike groups of the future through the timely, persistent air logistics missions our nation demands any place in the world. The CMV-22B is the Navy's long-range/medium-lift element of the intra-theater aerial logistics capability responsible for transporting personnel, mail and priority cargo from shore logistics sites to ships at sea.[2]

VISITING THE FIRST EAST COAST CMV-22B SQUADRON: OCTOBER 2024

I had a chance to visit Fleet Logistics Multi-Mission Squadron (VRM) 40 the "Mighty Bison" on Oct. 29, 2024. Earlier, I visited West Coast squadrons at North Island, San Diego, but this was my first opportunity to visit the squadron at Naval Station Norfolk.

This squadron will be a key part in supporting the fleet in the "contested logistics" environment now facing the U.S. Navy, a major challenge in both the Pacific and the Atlantic.

I had a chance to meet with the following officers: Commander Mason Fox, VRM-40 commanding officer; Commander Brett Learner, VRM-40 executive officer; Lieutenant Sam Ector, VRM-40 assistant operations officer; and Aviation Electrician's Mate Chief Petty Officer Frank Schaeffer, VRM-40 maintenance chief.[3]

We had a broad ranging discussion regarding the squadron and its preparation for its core missions.

Touring the squadron with Commander Fox.

During that conversation, Fox indicated how they met an unusual challenge for a new squadron. As they were getting ready to go from North Island (San Diego) to Norfolk on Dec. 6, 2023 with their first Ospreys, the DoD grounded the Osprey fleet DoD wide. Obviously, this was a shock but one which the squadron and its support community found a way to respond.

According to Commander Fox, the squadron had received their flight simulators so the pilots could train in the absence of flying real airplanes while waiting for the grounding to be lifted. And he indicated that the maintainers worked with Bell in Fort Worth on training the maintainers.

This meant that when the grounding was lifted in March

2024, the squadron was ready to re-commence their stand-up effort.

The core mission for the squadron is to replace the C-2A in the carrier re-supply mission. But because the Osprey can operate on a variety of ships, or from a variety of locations, it can provide for fleet support in a contested environment.

As Commander Fox put it: *The aircraft is very capable, and the pilots and air crewmen can do whatever mission we're tasked with for distributed maritime operation logistics. And that's the key point. If a flag officer says that they need to get a [supply] part to a submarine, we'll be able to do that.*

I've done so many different mission sets in my career, from ASW to attack to SOF support. All of them come down to time, distance, fuel and hover capability. If you can do time, distance, and fuel math calculations and understand your power margins, then really it's up to the people that task us with logistics to choose how they want to employ us.

What they have been focused on since the squadron has been activated is working with the carrier community on the logistics operations for East Coast based carriers, the Truman and the Ford. They worked with the Truman in June 2024 and the Ford in September 2024. They are planning to next continue their training with the Bush in the future.

The focus according to Fox: *We want to integrate as tightly as we can with the carriers and the air wings on the East Coast so that they fully understand our capabilities.*

They work with the CAG when onboard the carrier as does the C-2A. I noted when onboard the Ford that it has a significant design feature that will work well with the CMV-22B. The island on the Ford is not in the middle of the carrier flight deck but at the end which will allow the Osprey land on the carrier deck and be parked out of the way near the lift elevators onboard the carrier so not to get in the way of the launch and recovery operations going on the flight deck.

The first East Coast-assigned Navy tiltrotor vertical/short takeoff and landing (V/STOL) aircraft CMV-22B Osprey lands at Naval Station Norfolk, April 5. The CMV-22B Osprey belongs to Fleet Logistics Multi-Mission Squadron (VRM) 40 the "Mighty Bison." U.S. Navy photo by Mass Communication Specialist Seaman Sylvie Carafiol

The C-2A has several parts in common with the Hawkeye, but I asked how they will address parts issues onboard the carrier for the CMV-22B.

I was told that a footprint of support personnel will be set up on the carrier to deal with this need, somewhere in the vicinity of 15-20 people.

The CMV-22B is the Navy version of the Osprey but Fox discussed the importance of what I have called the tiltrotor enterprise for the joint fight and the contribution which Navy Ospreys can make as well.

According to Fox: *Our version of the Osprey has a little bit more gas that we can carry, and we have a primary mission that is different than the Marine Corps and Air Force Ospreys. But I think that if we're looking at a joint fight, we're looking at the 450 plus Ospreys that are part of the program record. They will all be contributing to distributed maritime ops, because that's the fight we are in.*

My final question was about how many aircraft are now in the squadron and what will be its eventual size. I was told that there were six planes currently in the squadron with a seventh to arrive the coming weeks. Dependent on final funding, they would have 12-15 Ospreys in the squadron.

Following our conversation, we all met in the squadron's temporary hangar. There is a new hangar being built nearby (two hangars down). The Ospreys rest already on the rebuilt tarmac next to the new hangar location.

THE PERSPECTIVE OF ADMIRAL VERISSIMO ON THE COMING OF THE CMV-22B TO THE ATLANTIC NAVAL AIR FORCE

I had the privilege of visiting Norfolk in the recent past and discussing the coming of the USS Gerald R. Ford with the first commander of Ford and then the commander of Naval Air Force Atlantic, Rear Admiral "Oscar" Meier.

Last month, I had the chance to visit Norfolk once again and to meet with Meier's successor, Rear Admiral Doug "V8" Verissimo and discuss with him the coming of the first squadron of CMV-22Bs to Norfolk and the evolution of the fleet in the Atlantic, which now includes the Ford carrier.

Naval Air Force Atlantic is the aviation Type Commander (TYCOM) for the United States Naval aviation units operating primarily in the Atlantic under United States Fleet Forces Command. AIRLANT is responsible for the material readiness, administration, training, and inspection of units/squadrons under their command, and for providing operationally ready air squadrons and aircraft carriers to the fleet.

Here the C-2A is seen parked next the Ford island and the weapons elevator is behind the photographer. This photo was shot by a Navy photographer when I was visiting the Ford on November 17, 2020.

Both the CMV-22B and the Ford carrier bring new capabilities to naval operations in the Atlantic region, and we discussed both during our time together on Oct. 29, 2024.

We started by discussing the challenge of contested logistics and how the coming of the CMV-22B provides significant capabilities to meet this challenge. Not only does the Navy need to deal with contested logistics, but consider this challenge in an environment where the Navy is focused on distributed operations.

There are benefits when the CMV-22 is combined with the Ford. The island on the Ford has been moved towards the end of the deck, freeing up space to which an Osprey can move when it lands for offloading of weapons or supplies, not blocking the EMALS catapults. And there is a fuel capability in that area of the deck which can refuel the Osprey for its departure from the deck as well.

Verissimo also correlated the coming of the Osprey with changes the Navy is working with in regard to its carriers. For example, he underscored: *The future will likely bring smaller more agile weapons to complement the heavier more difficult weapons to transport like TLAMs.*

He then argued that this shift to a different weapons stockpile would augment the utility of the CMV-22 supporting weapons re-supply in a contested combat environment.

He argued that there are specific capabilities of the CMV-22B which have a significant impact beyond logistics, namely, personnel support, notably in a medical emergency.

He put it this way: *If I have a medical emergency, I'm not trapping and catapulting the human body that's already injured. I can softly land and softly take off so I can take care of my people in a medical emergency.*

Throughout much of our discussion, the Admiral emphasized the evolution of the carrier for the new strategic situation and the flexibility it brings to the fight. The assets assigned to the Ford carrier, that contribute to the fight, will change as future payloads and platforms emerge.

He also underscored the unique features of the Ford design, notably the significant enhanced power generation capabilities which enable the ability to use future payloads, weapons and platforms which leverage that enhanced electrical power generation capacity.

The Admiral emphasized that the carrier brings unique capability to a blue water navy, and that the flexibility demonstrated through the life cycle of the Nimitz-class carrier and built into the Ford class is crucial for the fleet to adapt to evolving warfighting operations.

He argued: *The carrier and the carrier strike group is one of the only integrated forces which brings the core seven joint warfighting functions to the fight wherever it is operating. And with the Ford class, and its ability to generate electric power, it enhances those capabilities as well.*

12

SHAPING THE TILTROTOR SUSTAINMENT ENTERPRISE: PHASES OF DEVELOPMENT

Much like the evolution of the tiltrotor concepts of operations, the tiltrotor sustainment enterprise has gone through its own phases of development.

The evolution of the three variants of the Osprey have gone through three phases of development. The latest tiltrotor variant designed from the ground up for the U.S. Army, takes into account these lessons learned to set in motion an aircraft designed for sustainment in the digital twin age.

The first phase (2007-2014) of Osprey sustainment involved deploying the aircraft for operations in Iraq and Afghanistan. With the deployment began the real-world initial learning and support effort. The operators and the industry field service representatives (FSRs) worked together to support the aircraft in its various missions.

The Osprey is a very different aircraft than either fixed or rotary-wing, and it took time to learn how the operational characteristics affect the aircraft tip to tail in terms of maintainability. Learning about parts performance and shaping a supply chain that delivered parts in a timely fashion was a challenge to be worked.

The second phase (2014-2022) was shaping a regular maintenance regime rhythm. The aircraft is a digital aircraft that provides significant data with regard to its performance. The government worked with industry on performance-based logistics contracts that covered an increasing number of parts to be managed by the Bell-Boeing Original Equipment Manufacturer (OEM).

The third phase (2022-present) has been characterized by engineering redesigns of the aircraft to enable significant improvements in the sustainability of the aircraft leading to higher readiness rates.

Shaping the Tiltrotor Sustainment Enterprise

Phases of Development

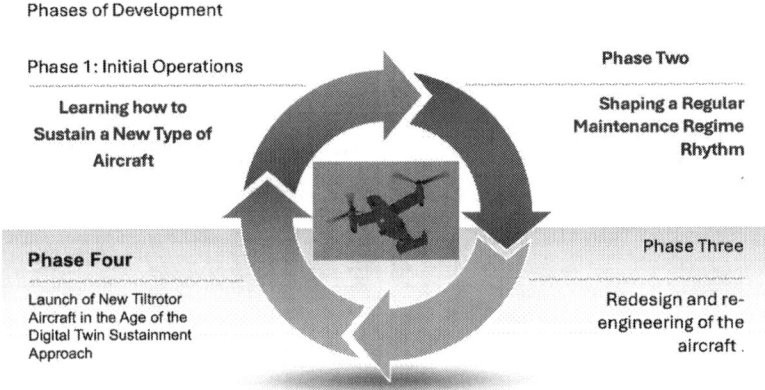

Phase 1: Initial Operations

Learning how to Sustain a New Type of Aircraft

Phase Two

Shaping a Regular Maintenance Regime Rhythm

Phase Four

Launch of New Tiltrotor Aircraft in the Age of the Digital Twin Sustainment Approach

Phase Three

Redesign and re-engineering of the aircraft .

As Brad Wanek, Director of V-22 Sustainment at Bell, put it in an interview:

Sustainment improves significantly with the knowledge and experience gained as the service operationalizes the aircraft. Once a level of operational maturity is reached, readiness levels begin to plateau, and it becomes more and more difficult to move the needle. To make significant improvements in readiness thereafter often requires major system improvements and hardware upgrades.

The first example of this has been the nacelle improvement program, introduced into the USAF fleet of Ospreys starting in 2022. The key objectives of the NI program have been to improve maintainability and reliability, so as to improve the V-22's mission-capable rate. A reduction in maintenance man-hours to troubleshoot and repair, coupled with reduced failure rate of components results in less downtime and increased mission-capable status.

A second example would be a major modification of the digital infrastructure of the aircraft, in which the computer power would be upgraded allowing for enhanced ability to handle and manage the data for sustaining the aircraft. This would enable other upgrades which rely on processing power and a transition to an open architecture system, as is being designed for the FLRAA.

This phasing is somewhat arbitrary, but it underscores the transition for learning how to sustain a revolutionary aircraft by taking those experiences in understanding how it operates and can be sustained more effectively and regularizing the knowledge and support system.

But one reaches a point where to craft major sustainment improvements on the aircraft, redesign and re-engineering of the aircraft is needed. This process of redesign and re-engineering is aided in part by the launch of the design process for the next aircraft in the tiltrotor enterprise, the FLRAA.

PHASE ONE: INITIAL OPERATIONS AND LEARNING HOW TO SUSTAIN A NEW TYPE OF AIRCRAFT

Standing up the Osprey in 2007 was a pioneer project. It was sent to Iraq and performed well in its first missions and was supported by the initial supply and maintenance approaches.

I first dealt with Osprey maintainers during a visit to USMC Air Station New River in August 2010. It was here that I began to understand that this was essentially the first digital aircraft that was generating data to aid in establishing a data-drive information support process to shape a more effective maintenance regime going forward.

A sense of the early days was provided by maintainers interviewed during that visit. An interview conducted during that visit provides insight into the challenges being faced in this first phase as perceived at that time.[1]

Question: What has been the challenge of moving from rotorcraft to a tiltrotor craft?

Sergeant Fante: "It was slow, really slow at first because mostly with lack of experience and lack of supply assets.

"Whereas, with the 46 and the 53, we see a part come in and if you didn't know what it was, you can easily go to QA (Quality Assurance), go to a senior guy to get help. Until we get a part in, everybody is kind of scratching their heads.

"We have to call a civilian an FSR, field service representative and they'd have to come in and kind of guide us along.

Sergeant Fante after the 2010 interview.

"The service vendors provided the transition, because we had no experienced Gunnys. The FSs here and they're very, very, very adequate.

"They're awesome guys but before with the traditional aircraft we relied on the long experience of Gunnys.

"And you have enough to pull in assets where you're going to have

parts and you have 30 year's experience producing parts so there's going to be a supply chain that's got capability built into it."

Comment: Whereas, in the case of the Osprey, the supply chain is new and you have challenges with availability of parts and metrics of performance and life cycle of those parts? So you're either going to have to adjust the metrics to make them realistic and you're certainly going to have to improve the product towards whatever the "normal" is.

That gets you back to your lack of experience issue. There's no gunny to go to and say, okay, we're going to work in sync for a long time, you know, I'm looking at this part, is this normal, right?

It's hard to have a normal, when you don't have normal.

Sergeant Fante: "We are dealing with a lot of new recognition issues with regard to the parts and their performance."

Corporal Warshek: "There is the problem of dealing with changing capabilities. Certain pieces of gear change almost constantly as upgrades are made. This meant we had to learn to adjust.

Corporal Warshek after the 2010 interview.

"For example, we just recently got in first two pendulum assemblies into our shop. We've never seen before. We rely on the FSR's to provide guidance on how we would tackle something like that. The pendulum assembly attaches to the rotor head."

Question: What was your experience in Afghanistan with regard to maintenance?

Corporal Warshek:"I deployed with 261, VMM-261 when they went out.

"Occasionally, we'd run in something like where would be in a situation where the service representative wasn't available or, we'd be on our own and the publications we were relying on was a little unclear or none of us had seen the problem before quite in the same way. So we would draw upon the "Osprey Nation."

A key challenge to effectiveness in maintaining the aircraft was the learning curve faced by the maintainers themselves.

I discussed the challenge of enhancing the knowledge and skill of the maintainers to take the sustainment enterprise forward in 2013 with Colonel Chris "Mongo" Seymour at his office in New River.

Marine Corps Air Station New River, N.C. - Col. Christopher C. Seymour, former commander of Marine Aircraft Group 26, thanks the Marines of his command for their hard work and dedication after his last flight, July 31. Seymour began his service as MAG-26 commander in May 2011 and is stepping down Friday, Aug. 2, 2013. Credit Photo: USMC

In a meeting I had with him during his last week of service at New River before retirement in 2013, he noted: *"There are several streams of activity which need to align to get the new system up and running and integrated into operations."* He spoke in detail about one *which was getting the Marines committed to owning the system and learning how to fix "new" problems, which come up with a new system.*

According to Seymour: "The challenges are different and must be worked differently. You need to get the maintainers to change their culture.

"Sorting out problems with the gearbox is a good example of what needed to be done. The gearbox on this airplane is very complex and central to its unique operational capabilities. The gearbox inside the nacelle turns a rotor, and they were chipping. This is high-end engineering. But when it was chipping, maintainers put it aside and waited for a new part. This meant the fleet was going to be degraded.

"The flight line needed to take ownership of the problem because a lot of it was self-inflicted. Maintainers would look to blame someone else when they had a proprotor gearbox go bad. As it turns out, the technology required was to use Hygroscopic oil that absorbs moisture out of the air, so if you have a gearbox that's not turning and boiling the water out on a regular basis, it goes long term down because of corrosion. It's sucking in the moisture of the North Carolina Coast into the oil."

PHASE TWO (2014-2022): SHAPING A REGULAR MAINTENANCE REGIME RHYTHM

As the maintenance practices became standardized, a Performance-Based Logistics System (PBL) set of contracts could be put in place which could more effectively manage the parts for the aircraft.

Defense Acquisition University defines a PBL as follows:

Effective PBL contracts contain core attributes necessary to deliver improved reliability and availability performance at lower cost.

These core attributes include:

- *A Performance Work Statement (PWS) that clearly defines the needed reliability and availability outcomes and value to the Department to achieve product support outcomes in unambiguous, objective and measurable terms*

- *The minimal set of metrics that directly support the stated outcomes, are specific enough to hold providers accountable, and do not inhibit or constrain the provider from achieving greater efficiency, cost control, incentivized productivity, and productive processes*
- *Incentives to deliver performance and reduce total cost*
- *A baseline and sufficient performance and cost insight to understand the price, cost, and "should cost."*
- *An understanding of the risks associated with non-performance and the strategies to mitigate adverse outcomes.*

To leverage PBL's proven ability to deliver needed reliability and availability to DoD customers at reduced cost, the DoD has a goal of ensuring the effective use of PBL, with particular focus on business arrangements that provide financial incentives to industry.[2]

The Osprey has seen several PBLs to date. Over time these PBLs have included more parts to be managed by the OEM but the Osprey has never had a tip-to-tail PBL which limits the ability of the OEM to manage the aircraft

Daniel Devereaux, the PBL program manager at Bell, in a 2024 interview with me identified the PBL path in terms of the inclusion of parts covered by these contracts.

The supply chain side of the PBL started in 2012. Currently, there are around 1000 depot level repairables right now: we have 228 of them. The government works with us on the 228 but goes directly to the vendors for the parts they make for the remainder.

With the U.S. Navy buying its variant of the Osprey and declaring IOC in 2022, an important stimulant to further change in the sustainment enterprise was provided.

I discussed this dynamic in the Fall of 2022 with "Mongo" Seymour, now with Bell and then head of sustainment for the Osprey at the company.

Recently, I turned to "Mongo" again to discuss the challenges and opportunities posed by the U.S. Navy now buying into the Osprey capa-

*bilities to support the carrier and the fleet. He is now in charge of sustain-
ment at Bell, and we picked back up from our earlier interview.*

He highlighted that while the Marines have significant experience in
sustaining the aircraft, the Navy has its own learning curve, just as did
the Marines earlier.

"It is sort of a 'ground-hog-day.' The Navy maintainers are genera-
tionally detached from the innovations of the past two decades of Marine
Corps operations and development. But they are inheriting more than 20
years of operational experience and are building from that forward with
specific U.S. Navy operational demands associated with the large deck
carrier and then with the fleet."

He pointed out that the Navy's acquisition of the Osprey meant the
Navy was taking ownership of the aircraft and making it a naval
aircraft instead of a unique Marine Corps platform.

Seymour pointed out that both services benefit from Navy ownership
of the platform; the Marines will benefit from Naval ownership and the
focus on support from the naval enterprise, while the Navy will benefit
from the established global sustainment capabilities and parts distribution points supporting USMC operations.[3]

PHASE THREE: RE-ENGINEERING AND REDESIGN
FOR ENHANCED SUSTAINABILITY

There is little question that the re-engineering and redesign
under way for the Osprey can significantly enhance readiness
rates, and that the Army-Bell partnership designing a new
member of the tiltrotor enterprise clearly builds on and
enhances this effort.

**The most notable example of this is the nacelle
improvement (NI) program.**

When discussing sustainment, an important metric is the
mission-capable rate (MC). That rate – expressed as a percentage
of total time an aircraft can fly and perform at least one mission
– is used to measure of the health and readiness of an aircraft
fleet.

The photo shows Albin in his military role as a CV-22 pilot.

The key objectives of the NI program have been to improve maintainability and reliability, so as to improve the V-22's mission-capable rate. A reduction in maintenance man-hours to troubleshoot and repair, coupled with reduced failure rate of components results in less downtime and increased mission-capable status.

I wanted to learn more about industry's role in the Nacelle Improvement program, so I turned to David Albin, the Nacelle Readiness Program Manager at Bell and interviewed him in February 2024.

For 20 years, David Albin served on active duty in the Air Force and in the New Mexico Air National Guard as a helicopter and tiltrotor instructor pilot, completing more than 200 combat sorties in the V-22 and rotorcraft.

From my experience with the U.S. Marine Corps in terms of the evolution of maintenance and sustainability, the first years were focused on getting the aircraft deployed to Iraq and Afghanistan and learning how to support it. The focus was on providing parts to ensure mission availability.

As a digital aircraft, the Osprey generated data on parts performance that allowed the Marines to understand better the maintenance profile of the aircraft. By the time it was anchoring the new Special Purpose-MAGTF or SP-MAGTF, Marines could make a reasoned judgment about what parts needed to be onboard the KC-130Js, which were flying with the Osprey on crisis management missions.

By 2015, enough data had been accumulated to focus on how to shape a sustainment enterprise. This enabled the Marines to achieve better mission-capable rates and lower cost per flight hour for the Osprey.

This is where nacelle improvement entered the narrative, as described by Albin: *When we started the nacelle improvement effort in 2014, we had access to data that allows industry to generate solutions using the fleet's data. We worked with government on the input from maintainers about the aircraft and looked for solutions to enhance the MC rate and lower cost for flight hour.*

Albin continued: *There are fixed costs and variable costs in working sustainment for an aircraft. We focused on the variable costs and how to reduce them.*

- *How do we reduce the demand for components?*
- *How many times are parts being ordered based on false positives from the diagnostic system?*
- *How do we reduce false positives or get more accurate reporting from the diagnostic system?*
- *How do we improve the choke points in maintenance which reduce MC rate, and drive-up cost?*

U.S. Air Force Airmen assigned to the 20th Special Operations Squadron familiarize themselves with the new nacelle improvement modifications on a CV-22 Osprey tilt-rotor aircraft at Cannon Air Force Base, N.M., Jan. 7, 2022. The improvements should increase aircraft availability.. (U.S. Air Force photo by Airman 1st Class Drew Cyburt)

The focus of the redesign effort was on engineering efforts to improve the operational characteristics of the nacelle.

Albin underscored: *The redesign focused both on service components to reduce the need for in-service repairs, like cracked frame stations, cracked baffles, the hinges and latches were all improved, so that maintainers would have to spend less time dealing with these components and their follow-on effects on the aircraft such as vibration in flight which caused the doors to open and potentially depart the aircraft, for example.*

He continued: *The Reliability & Maintainability Team used the data which had been accumulated from the operational fleet to determine what components or areas on the aircraft needed redesign. Based on this work, the engineers went and did the redesign and the NI program – then delivered reduced maintenance man hour rates and enhanced reliability.*

These combined effects of reduced maintenance man hours & improved reliability are what have holistically resulted in modified aircraft demonstrating higher MC rates compared to aircraft with the prior variant of nacelle.

This approach which yielded the NI program was rooted in two things: (1) a demand side shaped by the maintainers, and (2) the data generated by the aircraft from the operational fleet.

This information then flowed into industry, which then could parse the data and convert the maintainer input into engineering requirements. The engineers then focused on specific, realistic solutions in a re-design precisely focused on a more sustainable aircraft.

The resulting program had four key lines of effort:

- New build of the nacelles;
- Enhanced reuse of repairable components;
- A new wiring design which improved maintainability, reliability and reduced part count;
- New structure, consisting of targeted improvements to address fleet needs.

What has been the result?

According to data through the end of 2023 from the first users of the NI effort, namely the Air Force CV-22 community, the results have been significant. Twenty-one NI modified aircraft have flown 4,065 hours to date. During those flight hours, the maintainability improvements of NI have saved over 10,000 maintenance man hours or over 400 days of maintainer time on the modified aircraft compared to the time that would have spent on the legacy nacelle design.

Based on the NI program objective to improve reliability by four times, the prediction for NI after over 4K flight hours was 140 component failures. The actual failure rate of NI components to date has been zero, which is a truly significant result.

With regard to the NI maintainability rate, the results have also been notable. The objective was to reduce maintenance man hours by 75%, which after 4K flight hours should have accrued 2,195 hours. The actual accrued maintenance man hours on NI components are at 12-man hours to date, which is a remarkable outcome.

As for the MC rate benefit from NI, in October 2023 AFSOC observed a 10.8% MC rate improvement in their NI modified aircraft compared to the legacy nacelle aircraft in their fleet.

Industry predictions are for an overall MC rate improvement of 7% or higher for the CV-22 fleet once all 50 aircraft in the Air Force fleet are modified.

The data leads to one conclusion: The NI program is a significant step forward in shaping a more sustainable tiltrotor enterprise. The benefits to the fleet from improved maintainability, reliability and overall MC rate are certain to provide great benefit in the austere and distributed operations employment of the aircraft in the future.

NI is suggestive of how reengineering the aircraft can significantly expand sustainability and its operational life. In an article published on June 10, 2024 by Jo Ann Y. Williams, the author cited the perspective of Col Brian Taylor, the V-22 program manager at NAVAIR regarding approaches they are either working or considering at NAVAIR:

A variety of configurations presents a host of challenges for aircraft maintenance, for supply chain management, and for the introduction of new capabilities.

To address this, Col. Taylor reported that the Air Force, Marine Corps and Navy are "coalescing on a kind of standard configuration, which is huge." Huge, indeed!

However, while the services strive for what they consider to be a standard configuration, it's more important for them to strive towards an "operationally relevant" configuration management plan.

Col. Taylor used a helpful metaphor in thinking about this. "If you think about the car that you bought 25 years ago and the car that you bought today, they are very different. They have different systems, and so [we're] trying to kind of normalize all the systems and everything on an aircraft."

A prime example is the Osprey's mission computer system. Col. Taylor explained that because there are two variants of the Osprey's mission computers, two different software builds are required. That's expensive, complex, and time-consuming. Shifting to a single baseline would save taxpayers money and provide more capability, more rapidly, and that improves the experience for V-22 pilots as well.

Col. Taylor noted that "not having the software dependent on hardware" provides "a lot more flexibility in the fleet for mission kitting and things like that."

Williams then further quoted Col Taylor as follows: Then there's the need for a cockpit refresh. Col. Taylor used a helpful metaphor – an older-model car. When you drive a new car off the lot, typically the dash has touchscreens; it interfaces with your mobile devices. A car that's 20 years old has none of that. And, everyone knows it's more expensive to replace 20-year-old parts!

"These are a bunch of screens and displays and keyboards and stuff that were developed, back in the late 80s, so keeping them on the aircraft is pretty challenging," he said. "We are kind of at a tipping point where we are spending enough on just maintaining what we have that it's time to do something different."

The answer from the Joint Program Office is a program called the V-22 Cockpit Technology Replacement, or VeCToR. Col. Taylor noted that commercial, off-the-shelf technologies will be a big part of the solution.

The Open System Architecture (OSA) should be a priority, as it can evolve and adapt as future threats emerge.

Leveraging work on the Army's FLRAA program would provide the Navy and Marine Corps the ability to address legacy system constraints such as computer processing and display interface, while pro-

viding a significant cost savings and risk reduction and providing a path to interoperability with the U.S. Army, other services, and foreign militaries who adopt FLRAA variants.[4]

In addition, the following upgrades are in planning and implementation phases for the purpose of increasing the safety and reliability of the V-22:

- Osprey Drive System Safety and Health Instrumentation (ODSSHI): A vibration sensing system upgrade will be implemented to identify components that need to be replaced prior to in-flight faults. Testing is currently underway.
- Proprotor Gearbox (PRGB) pinion bearing redesign: The updated design incorporates new materiel into the pinion bearing for safety and reliability. This improvement is awaiting production and installation.
- V-22 improved Input Quill Assembly (IQA): An improved IQA design based on Hard Clutch Engagement (HCE) investigations will be updated. The prototype design is entering the qualification testing phase.

As LtGen (Retired) Heckl put in the forward to this book: *The Marine Corps will fly the Osprey for at least 50 years, to the year 2060 or most likely well beyond. I hope Robbin's book will contribute to the acknowledgement that the MV-22 has been an absolute success, and it is time, as we approach the 20th anniversary of the aircraft's IOC, that we look to make the investment to ensure this one of kind capability, stays healthy and the community robust.*

❦ 13 ❦

THE OSPREY
MAINTAINABILITY AND
SUSTAINMENT ECO-SYSTEM

T he Osprey was built from the outset as a digital aircraft and a pioneer in the introduction of data-rich systems for maintainability and sustainment. The V-22, as is the case with the F-35 and CH-53K, is a very capable and complex aircraft that has needed attention to detail to make the entire enterprise work as part of a comprehensive sustainment system.

Readiness considerations in public discussions of the aircraft have tended to focus on the availability of parts, certainly a consideration, but understanding the entire eco-system is the key to generating higher readiness rates.

How might one characterize the Osprey's maintenance and sustainability eco-system?

Brad Wanek, Director, V-22 Sustainment at Bell provided this definition: *I see three major components of the V-22 eco-system related to aircraft readiness. These operate in concert and must be in proper balance to deliver the optimal outcome.*

Each of these three are interdependent subsystems, which cannot operate without inputs from the other two. They are supply, maintenance, and information.

The supply system is the most objectively quantifiable and, as a result, the easiest to measure and track performance. Because of the readily available metrics, supply has, in my opinion, received disproportionate focus of effort.

Maintenance system performance is as much or more a driver of aircraft readiness as supply, but it receives less focus. We can track maintenance management effectiveness overall, but the subcomponents of maintenance capacity and capability, such as maintenance training quality and the experience and proficiency of the maintainers, themselves, are more difficult to measure. Accordingly, when it comes down to metrics, it receives less leadership attention.

Information flow is critical to the functioning of both supply and maintenance, yet it is relatively neglected in discussions about fleet readiness. The quality and capability of the information systems available to manage the maintenance and supply systems are critical factors in aircraft readiness.

Having the information systems which can harvest knowledge from the aircraft to inform maintenance and supply personnel and having the capacity to utilize that knowledge to shape and improve maintenance and supply methods and procedures will ultimately determine the level or readiness that can be achieved.

In this very thoughtful characterization, we see a dynamic interactive system among physical parts (supply), the level of performance of maintainers and their support element and finally, the increasing importance of exploiting the capabilities of data rich aircraft like the V-22.

The parts production issue is a key one as supply chains post-pandemic have been taut. It is also the case that governments in the West have followed a just-in-time parts production and delivery system which has guaranteed there is not a deep supply of production capacity.

The war in Ukraine has provided a stark reminder of how challenging it can be to ensure a robust supply chain response in the event of an unanticipated evolving crisis.

A Marine, with Marine Medium Tiltrotor Squadron 365
(Reinforced) (VMM 365), performs maintenance on an MV-
22B Osprey, assigned to the Blue Knights of (VMM) 365
(Reinforced) on the flight deck of the amphibious assault ship
USS Bataan (LHD 5), March 18, 2020. Â Photo by Seaman
Darren Newell. USS Bataan (LHD 5)

Then there is the question of the arrival of 3D printing. It is now possible to produce parts at the point of operation thereby changing the mix of prefabricated parts which need to be shipped to the point of operation and those which can be produced locally. This obviously is a work in progress and the proper mix will be determined on a case-by-case basis.

Efficient delivery of supplies is a key element as well. With Marines emphasizing expeditionary forward basing, the question of where to position supplies to ensure readiness at the tactical edge is a key consideration. It is not simply stockpiling at fixed points; it is about availability of parts to sustain the tip of the spear operations.

The second element for eco-system consideration is the experience and proficiency of the maintainers themselves. I have interviewed many maintainers of the Osprey over the years, and the evolution of how they are trained, how they work with industry service representatives, and how they share knowledge is a key issue.

The place of origin for Osprey operations, 2nd Marine

177

Aircraft Wing, faced a difficult logistics challenge from the very beginning because squadrons were required to deploy into combat soon after Initial Operational Capability (IOC) was attained. In many cases, maintenance practices were learned on deployment. This is a problem the Marine Corps plans to avoid with standup of the new version of the CH-53 by delaying until 2025 or 2026.[1]

Given the difference of the V-22 from any air system that proceeded it, cultural change was crucial as well. That point was made during a 2013 visit to MCAS New River and my discussion with Colonel Chris "Mongo" Seymour.

Getting the Marines to own the system and learn how to fix new problems, which always come up with a new aircraft was essential. The problems are different, and they must be worked differently. You need to challenge the maintainers to change their culture.

Over time, the role industry service representatives in helping active-duty maintainers has ebbed and flowed, although with the introduction of the Osprey into the U.S. Navy fleet, their role has been a central one in shaping the hands-on training approach.

But there remains a fiscal challenge as well as policy questions regarding the role such senior industry personnel should play in the future.

Personally, I think the role is a significant one as the U.S. military evolves and the recruitment and re-enlistment challenges seem to be a steady challenge facing the force. A complex aircraft needs an experienced corps of maintainers.

The third key element of the interactive sustainment eco system is information and knowledge obtainable from the data generated from the aircraft and the maintenance regime.

John Russell, Manager, V-22 Field Services of Bell, characterized the information piece as follows:

The aircraft has multiple components and sensors that monitor and record flight profiles, system functionality, vibration analysis, and exceedances.

At the end of a flight this data is extracted from the aircraft and uploaded into a computer system that displays the information for maintainers to review.

This system, known as CAMEO (Comprehensive Automated Maintenance Environment - Optimized), provides fault and exceedance information for maintainers to evaluate. The maintainers need to have an expansive understanding of aircraft sub-systems, theory of operation, and maintenance related tasksÂ in order to interpret the data properly.

If faults, failures, or exceedances are identified, CAMEO will provide trend analysis and general references for corrective actions.

If engineering disposition is required, more detailed data can be extracted for evaluation.

All CAMEO data, fleetwide, is uploaded to the government's Readiness Integration Center to provide Automated Logistics Environment (ALE) support to engineering, logistics, and maintainer communities.

The Osprey has been a pioneer program in terms of generating such data for the services, the government and industry working together to leverage the data to gain better insight into the performance of the aircraft and its parts.

But to maintain expertly the aircraft, one needs to know the overall system of the aircraft and not simply be a specialist on a single subsystem of the aircraft.

It is clear that the three elements of the Osprey's sustainment eco-system, as described here, are highly interactive.

The physical supply side is affected by the actual performance of the parts, the ability to move the parts to the point of maintenance, and to the manufacturer being able to improve parts through input from operational experience.

The performance of the maintainers decisively affects aircraft readiness, and their collective knowledge provides feedback throughout the system and provides an important input to the manufacturer regarding parts improvements.

Industry Service Representatives can play an important role in providing information back to the manufacturer regarding improvements in parts and procedures as well.

The information generated by the aircraft and flowing through the logistical system forms a critical digital backbone for understanding the aircraft and informing the process of product improvement.

Over the course of almost 500,000 flight hours, the Osprey's maintenance and sustainability eco-system has become ever more robust and effective.

These lessons learned will continue to evolve and ultimately set the stage for improved readiness and reliability of this aircraft as well as future advanced tilt-rotor variants now being developed.

14

SHAPING THE FUTURE OF
THE ENTERPRISE

When the Marines brought the Osprey to Iraq in 2007, they were introducing a revolutionary aircraft which change how the ground combat element could be deployed in the country. The plane could cover any part of the country within two hours and simply blowing past the time-space considerations of a rotorcraft.

There was an especially poignant story told to me in an interview I did in 2010 at UMSC Air Station New River.

According to Major Lee York in that interview: *"We took some soldiers out to the West of Iraq. The crew chief comes up to us and tells us that the guys won't get out of the plane. We're like, :what are you talking about? They said we're not there yet."*

And we said, "What are you talking about?"

He then said, "The last time we did this flight it took an hour and a half. We've only been in the plane for 40 minutes so we can't be there yet."

We told him to tell the Marines that "we were cruising at 230 rather than at 120 so we were there. I swear we're here, you know, we're not going to send him somewhere where he is not supposed to be."[1]

That was the beginning of a journey since 2007 whereby the aircraft led to the Marines, the Air Force and now the Navy learning how it allows them to operate completely differently from a rotorcraft on the chessboard of operations.

Now the U.S. Army is going to inherit this experience and leverage it as they build their new variant of the tiltrotor aircraft, the Future Long Range Assault Aircraft or the FLRAA.

But their journey begins at a different launch point. Not only can they drawn on the operational experiences since 2007, but they are designing the aircraft in the evolving digital age and building a solid foundation for shaping from the outset a collaborative sustainment enterprise between industry and the Army which will allow for more cost-effective aircraft readiness rates so to speak by design.

And one is discussing sustainability from the design stage one is talking about working how to build into the aircraft tools which allow Army maintainers to work the aircraft from the outset in ways to redesign how they will maintain this aircraft.

The program is focused on future vertical lift and the key element about shaping the future aircraft is to combine the performance characteristics of the tiltrotor aircraft with a new sustainment approach built into the aircraft. The design to do so is not just the aircraft but the overall configuration of the sustainment approach.

The FLRAA draws upon the experience of the tiltrotor enterprise to date in terms of being focused on the performance characteristics which tiltrotor can deliver to the Army. Illustrative of how the Army is thinking about FLRAA and its impact on Army operations is the perspective on how Army Special Forces intends to use the aircraft.

According to a *Flight International* article published on May 8, 2024: *Rotary-wing procurement officials at SOCOM confirm that the final V-280 design will incorporate features allowing it to be altered quickly and at minimal cost with special operations features like an air-to-air refuelling probe and nose-mounted terrain-following radar.*

"When it does come to us, it's a vastly reduced effort," says Keough.

"We're still going to monster-garage it, but now it's going to be a lot more affordable...."[2]

And another source indicated that the SOCOM requirements were being worked into the FLRAA design process which allowed SOCOM investments to be made as well in the aircraft.

According to a May 13, 2024 article by Michael Marrow:

SOCOM's Rotary Wing Program Executive Officer Steven Smith first revealed that the PDR had occurred during May 7 comments at the SOF Week conference, and said the command and the Army have had a years-long dialogue to "influence the FLRAA design," which yielded changes to the aircraft as a result.

"The good news is prior to the recent completion of their PDR, that they've adopted all of our changes to the aircraft," Smith said, according to Aviation Week.Per Smith, that included specific nose design requirements, hardware for a refueling probe and "a couple other minor things that we've asked for to get into that platform that will make our modifications less expensive down the road."

The collaboration, Smith said, is a "a good news story" that will make it "easier for us to incorporate all the secret sauce, all those boxes that we put on the aircraft that provide our unique capability."[3]

In other words, tiltrotor will deliver significant high-performance capabilities to the Army as it has to the other users of tiltrotor. It can move in the battlespace rapidly and at distance and surprise the enemy from a diversity of locations, or one can disaggregate the force and can exercise such flexibility because one can do it all on the same day.

Rebecca Grant wrote about the role of FLRAA in a broader context of the shift in Army concepts of operations in a June 4, 2024 *Breaking Defense* piece. She concluded her piece as follows:

Developing the new equipment and concepts for air assault over thousands of miles is an enormous challenge, just as it was in World War II. The U.S. needs strategic agility in the Pacific islands, and the Army's V-280 Valor will have to be a big part of it.[4]

But she wisely underscored in the piece that the new tiltrotor aircraft will need to be part of a strategic redesign of the

Army for it to be used properly. Much as the Marines in pioneering the Osprey revolution reworked their concepts of operations throughout the past years, the Army will face a similar challenge. And the USMC as it positions itself for adding FLRAA to its expeditionary force, might well be able to help shape Army concepts of operations as well.

Grant underscored the importance of FLRAA and new concepts of operations to synergistically develop together .

Longer range air assault will enable ground forces "to converge through decentralized operations at extended distances." Basically, the idea is to box in China by placing forces on islands or terrain features. Rapidly inserting forces can extend US sensor coverage and create options for positioning land-based fires.

Crucially, the U.S. can also marshal forces outside China's missile threat rings. For example, the V-280 Valor, sustaining at 240 knots, can deploy from Hawaii to the Philippines in 20 hours. Down the road, the V-280s may also be able to self-deploy, moving to the theater as unmanned aircraft and rejoining with crews at a forward location.[5]

But future vertical lift is also about taking this performance capability but lowering the burden of employment and lowering the burden of sustainment. There is a focus from the outset on lowering the burden on producing sorties.

How is the program focused on doing this?

The core principle is upon how to harmonize the sustainment construct with the core characteristics of the aircraft.

One focus is upon leveraging data from a fly by wire aircraft which is self-aware and constantly reports on itself is a key element. The Army has a well-developed enterprise management IT system for its aircrafts, and there is a focus on designing the aircraft to work with the right kind of data management systems operated by the Army to deliver the desired sustainment enterprise.

The design process is being worked in the era of the digital twin manufacturing process. Such a process allows for effective production of a common aircraft which enhances effective sustainability. And the IP in the digital twin can be transferred to 3D printers allowing the use of 3D printing in the field to enhance a new sustainment process.

A re-designed tiltrotor taking the experience of decades of Osprey experience into the design process from the outset and in a digital twin era will lead to a much cost-effective and sustainable tiltrotor aircraft. But the Army variant is smaller and has features not found on the Osprey which means that it is indeed an enterprise yielding aircraft with different capabilities and characteristics.

The Army announced in early August 2024 that the new tiltrotor project had moved ahead to Milestone B with the expectation that the first FLRAA flight will be in 2026 with Low-Rate Initial Production scheduled to begin in 2028 and initial fielding activity in 2030.

Redstone Arsenal, Ala. – The Army's Future Vertical Lift program took a major step forward as the Future Long Range Assault Aircraft (FLRAA) program entered the next major phase of development when the Army approved the FLRAA Milestone B (MS B) Acquisition Decision Memorandum this week.

The decision came after the successful FLRAA preliminary design review in April and a meeting of the Army Systems Acquisition Review Council (ASARC) in June. After reviewing FLRAA affordability, technological viability, threat projections and security, engineering, manufacturing, sustainment, and cost risks, the ASARC confirmed that all sources of program risk have been adequately addressed for this phase of the program. MS B allows the Army to exercise contract options and continues development of the aircraft as it now enters the Engineering and Manufacturing Development Phase.

"This an important step for FLRAA and demonstrates the Army's commitment to our highest aviation modernization priority," said the Army Acquisition Executive, the Honorable Douglas R. Bush. "FLRAA will provide assault and MEDEVAC capabilities for the future Army, adding significantly increased speed, range, and endurance."

"This is an exciting day for the Army... and more importantly for our

Soldiers. The FLRAA provides truly transformational capability to Army Aviators as we uphold the Sacred Trust with the Soldier on the ground," said Maj. Gen. Michael C. McCurry, 17th Chief of the U.S. Army Aviation Branch. "Future battlefields require expanded maneuver, the ability to sustain and provide command and control across vast distances, and of course, evacuate our wounded. All of these apply to both conventional and Special Operations Forces. With roughly twice the range and twice the speed, FLRAA brings unmatched combat capability to the Joint Force."

The Army awarded the FLRAA Weapon System Development contract to Bell Textron on 5 December 2022 and it includes nine options. The Milestone B allows the Army to exercise the first option which includes detailed aircraft design and build of six prototype aircraft. The Army is planning for the first FLRAA flight in 2026 with Low-Rate Initial Production scheduled to begin in 2028 and initial fielding activity in 2030. The Army will continue to review and refine the schedule as necessary based on the contract award and the latest program activities.

"PM FLRAA and our Team of Teams across the aviation enterprise are working hard to make sure that we get it right," said Brig. Gen. David Phillips, Program Executive Officer, Aviation. "We will deliver a next generation combat capability that meets the Army's goals for affordability, survivability, maintainability, reliability, and safety."

"The FLRAA Milestone B decision is another successful step of a deliberate modernization effort by the Army," said BG Cain Baker, Director for the Future Vertical Lift Cross Functional Team. "The many stakeholders, including academia and industry, have worked hard to ensure rigorous technology development and demonstration and have informed FLRAA requirements and affordability. FLRAA's speed, reach, and survivability will be key to transforming US Army maneuver."

"I am very proud of the FLRAA team. We've maintained a deliberate balance between sustaining program momentum while maintaining technical and acquisition rigor," said Col. Jeffrey Poquette, FLRAA Project Manager. "Using digital engineering as a key part of our 'go slow

to go fast' approach has helped to accelerate the program by investing in requirements development up front."

FLRAA will provide transformational capability for ground forces and aircrews with speed, range, and surprise to present multiple dilemmas to the enemy. It will expand the depth of the battlefield, extending reach to conduct air assault missions from relative sanctuary while enabling us to rapidly exploit freedom of maneuver to converge ground forces through decentralized operations at extended distances. FLRAA's inherent reach and standoff capabilities will ensure mission success through tactical maneuver at operational and strategic distances.

As the Army transforms to meet an uncertain future, FLRAA is one of many modernized capabilities that will help ensure the Army of 2030 and beyond is ready to win when the nation calls.[6]

❦ 15 ❧

THE WAY AHEAD WITH FLRAA: INSIGHTS FROM AUSA 2024

T he panel on future of vertical lift and interviews conducted by the press during the October 2024 Association of the U.S. Army (AUSA) conference provided further insights regarding the FLRAA program and the way ahead. The conference was held from the 14-16 October in Washington D.C.

Let us address first the schedule indicated by these sources plus information associated with achieving Milestone B. Milestone B was achieved in the summer of 2024 as described in the previous chapter.

The next phase was described by one source after the AUSA conference as follows: *Bell is to deliver two virtual FLRAA prototypes – exact digital replications of what Bell intends to build to the army in February 2025. One copy is to be sent to Fort Novosel, where the army's Aviation Center of Excellence will examine the design. Another copy is headed to Redstone Arsenal for analysis by Army Materiel Command. The virtual prototypes are also intended to fulfil a Middle Tier Acquisition requirement to deliver working prototypes to the field before transitioning to a programme of record, which cannot be done with a complex aircraft.*[1]

Another source added: *A so-called Critical Design*

Review (CDR) is expected to occur next summer ahead of the planned delivery of the first of six FLRAA prototypes in 2026. The program schedule as it exists now sees low-rate initial production of the tiltrotors starting in 2028 and units beginning to use them operationally in 2030. [2]

What underlies progress to the next phase is the design and development approach being followed by Bell and the Army with regard to the platform. Digital engineering is how the platform is being design which allows for a more rapid and effective development process which changes how requirements are shaped in the process of the build of the aircraft.

The panel on the future of vertical life explicitly discussed this process. Colonel Jeffrey Poquette, project manager for the future long-range assault aircraft described the approach as follows:

So I'll start with the FLRAA specifics, because as General Schlosser said, we are kind of the pilot case for digital engineering. We like to say that the program is born digital — a cleansheet design, the first time it's been done in forever. We get a lot of extra homework from the Department of Defense, the Army, and even the GAO to talk about the benefits of digital engineering.

Digital engineering is a large umbrella; there's a lot that falls under it. Model-based systems engineering falls under that umbrella. I'll focus on the digital environment.

The digital environment is a collaborative workspace. Let me talk about how we used to do Previously, I would hand over words in the requirements document and expect Bell to go do some things and send me some engineering artifacts back. They would spend months doing that, and we would spend months reviewing it. If we found something wrong, we'd send it back, and we would iterate —that takes time.

In the digital environment, we are working in real time with Bell every day, looking at what they're doing, and we are course-correcting them as needed. We catch issues right after the Preliminary Design Review (PDR). When they submitted their model of the system after PDR, we turned around about 3,000 things we needed them to address within two weeks.

These are not big things; they are just items to ensure we get it right. I want to make clear that that's the goodness of digital engineering — finding those issues now and ensuring we don't build the wrong thing is a testament to the power of digital engineering.

So you asked how it makes for a smooth program. I'm not sure if it makes it smooth; what it does is make it faster. It allows us to iterate quicker and collaborate in a way that was just unheard of in the past. Another thing that we do, like digital model-based systems engineering, is that it's all done in the model. The likelihood of missing a requirement that General Baker handed to me is almost zero because it's not just words on a page; it's in the systems engineering model.

So we're really proud of it. We talk a lot about it, and the Army is going to continue to emulate the things that we've done, get lessons learned, and develop weapon systems this way into the future.

Major General Michael McCurry, Chief of Staff Futures Command added: *You heard yesterday our chief in his address that's now on lunch say, "Step on the gas; we've got to do things faster." He also talked about process innovation. Well, this is an example of process innovation, and we have to continue to innovate our processes inside the Army as well.*

Dr. James Kirsch, Director of the Army's Aviation and Missile Center added: *I'll add a little more to that, because we talk a lot about going fast, and a lot of times we're talking about going fast in the design and development phase. But there's another piece of that before we can actually hand kits to soldiers, and that's getting through the material release process, which includes airworthiness for aircraft as well. That generally happens at the end. We get a lot of documents and a lot of data from the PM as they're finishing up their product, and then we spend a lot of time evaluating and determining whether or not we think this aircraft is safe to fly.*

Now, in this digital environment, we're involved in the process from the very beginning, and we can watch those things develop over time. We have much more access earlier in the program to the data, which should help us get to those pieces of the process — the airworthiness releases, the safety releases, and the material release process—that actually allow us to get that kit they've developed into the hands of our fighters that much quicker.

Digital engineering allows the delivery as mentioned about of a virtual prototype in 2026 as a key deliverable.

Colonel Jeffrey Poquette discussed this aspect of the program as follows: *The other thing I'd like to highlight is that this is a kind of a first for aviation, as we're delivering a virtual prototype... With that virtual prototype, we'll put one down at Fort Novosel and one up in Redstone. It will inform doctrine and training, expose Army aviators to tilt-rotor technology — which they haven't experienced before — and it will be a design tool for Bell and the program office to iterate on and get to the point where we can fly. So, the first prototype delivers in 2026, and flying will occur sometime shortly after that.*

A key aspect of the program is engagement by the operational force in becoming familiar with the platform and how that platform will impact Army concepts of operations.

Familiarization was highlighted by Colonel Poquette during the panel discussion.

We're located in Huntsville, AL. My initial thought was to let's go to the 101st; they're right up the road. I had a conversation with General Baker, a former 25th commander and deputy commanding general there. He said, "Why don't we go to the place that we really think this aircraft matters?" It was a perfect idea. We executed it, and it took an extra month to get all the logistics in place. We sent the Bell team and my human systems team out there to bring the mock-up to receive input from the 25th.

The 25th CAB took place with the guys who could potentially fly it, and the combat brigades out there came across. We spent a week out there with them, iterating on the two configurations we wanted to test. We started with one configuration that we thought we were going to choose, with the seat configuration. We ran the soldiers through, you know, without any combat gear. Then we added weapons, then we added body armor, and then we added MOPP gear.

We timed them. The people that do this kind of human systems science are technical experts. We measured every soldier with giant human-sized calipers. We wanted to ensure we could take care of the 5th percentile soldier and the 95th percentile soldier. We did that in two configurations. What we found out is that the soldiers really liked the second configuration better.

Now that they liked it, why did they like it? It was more comfortable, it was faster, and they weren't tripping over each other. We got feedback on the cockpit; the cockpit was mocked up with a lot of 3D-printed controls and that kind of thing. The weapon was different. When we were in location, we had the crew chief seat oriented differently.

So we took all this into account and we ran that combat speed. Even as the soldiers came out, they ran out and went into the prone with their weapon. It was as tactical as we could make it while getting the information that we needed.

We then take the soldiers after they spend a week doing this, and I will say this: it sounds super cool, but it's not a lot of fun for the soldiers. I mean, being in MOPP gear and body armor, getting on and off the aircraft multiple times — it's challenging. So I made a very special point to pull all these soldiers in and said, "Look, I know this isn't fun, but the information and insight that you are providing to us is invaluable, and it will be incorporated. You are part of something historic — Army acquisition history for Army aviation. One day, when you fly this, or your children fly this, or you see this flying above you while you're on your front porch as a grandfather or grandmother, you're going to know that you had input into this aircraft." That really resonated with them.

And the Army is working the impact on con-ops challenge now as well in advance of receiving an operational aircraft.

As Audrey Decker noted in *Defense One* piece published on 16 October 2024: *As the Army awaits the arrival of its high-speed, high-capacity tiltrotors in 2030, it is already practicing the new operating concept that they will enable. The idea behind "large-scale, long-range air assault," or L2A2, is to "deliver one brigade combat team in one period of darkness, over 500 miles, arriving behind enemy lines, and able to conduct sustained combat operations," said Maj. Gen. Brett Sylvia, who leads the 101st Airborne Division. And the service can't do this with the platforms it has today, he said.*

Sylvia's team has started working on the new tactics, techniques, and procedures and has practiced the concept four times in live demos over the last year, and multiple times in simulation. In the simulation, replacing UH-60s with FLRAA gave the combat aviation brigade "four times the amount of heavy-lift aircraft than what I have today," Sylvia said.

In the most recent test, the unit used its existing helicopters to move a brigade combat team about 570 miles, from Fort Campbell on the Kentucky-Tennessee border to Fort Johnson in Louisiana. The movement required three nights, two mission support sites, and six forward arming-and-refueling points. But in a simulation that used FLRAAs, the same mission required half the sustainment and security footprint--and just one night, Sylvia said.

"We are building, over the course of the next few years, this air assault combat aviation brigade. We are doing the things in order to be able to build the foundation so that all we have to do is just receive the aircraft and we'll be ready to execute," he said.[3]

16

LAUNCHING A NEW MANNED AIR SYSTEM AT THE DAWN OF AN AGE OF AUTONOMOUS SYSTEMS

T he Future Long-Range Assault Aircraft is being designed to operate in a new world of combat, namely one in which autonomous systems will become significant players.

I would argue indeed that when introducing new manned systems now and in the future, it is increasingly important to do so with consideration of how they can work offensive and defensive operations in a world where autonomous systems will become ever more prevalent and prominent.

What is impressive about the Army's standup of their new tiltrotor aircraft is how they are doing so with core consideration of how the concepts of operations of assault operations will change with the combined arms approach which is inherent in working with and defending against autonomous systems.

A panel held at the October 2024 Association of the United States Army convention held in Washington DC. discussed the way ahead with FLRAA. This panel was hosted by *Defense News* which provided a video of the panel which provides readers the opportunity to watch the discussion by senior Army leaders.[1]

The members of the panel were:

- Major General Michael McCurry: Chief of Staff Futures Command;
- Brigadier General Cain Baker: Director of Future Vertical Lift Cross-Functional Team;
- Colonel Jeffrey Poquette: Project Manager for the Future Long-Range Assault Aircraft (FLRAA);
- Dr. James Kirsch: Director of Combat Capabilities Expand Aviation and Missile Center.

The panel started with a discussion of the changing military context into which the FLRAA is being introduced. The conflict in Ukraine highlights the rapid pace of technological advancement and the need for agile and adaptable military capabilities. This necessitates a shift away from traditional platform-centric thinking towards a system-of-systems approach, prioritizing formations of capabilities.

The FLRAA program is a significant investment aimed at transforming Army aviation. It prioritizes speed and range and leveraging digital engineering for rapid design and development. Soldier feedback is actively integrated into the design process, ensuring the aircraft meets operational needs.

Yet at the same time, the Army recognizes the crucial role of unmanned and autonomous systems, particularly launched effects (ULEs), in future warfare. These systems enhance situational awareness, provide stand-off capabilities, and contribute to holistic team survivability. Ongoing experimentation and exercises like the Army Futures Command's (AFC) Future Vertical Lift (FVL) Cross-Functional Team (CFT)'s Experimental Demonstration Gateway Event (EDGE) and Project Convergence are informing doctrine, organization, and training for effective ULE employment.[2]

But how to do both, launch a new manned platform and integrate ULEs?

According to Major General Michael McCurry: "We aim to capitalize on the strengths of both without sacrificing humans for first contact, focusing on what machines can do best and what humans do best. The most important letter in HMI is the "F" for formation. We're focused on formations of capability, which is a bit different from others around the world that want to employ a singular material item on the battlefield. We're interested in building formations and capabilities at echelon. A great mentor, retired General David Perkins, once told me to quit focusing on one thing and see how it fits into the bigger picture. So, this formation-based approach is critical."

In other words, the focus is upon leveraging what a new platform like the FLRAA can provide, namely speed and range, but working in what I have called combat clusters to leverage what autonomous systems can deliver, or in the words of the panel how ULE employment working with FLRAA shapes the concepts of operations of operational units in the future force.

As Colonel Jeffrey Poquette underscored: "Two things about FLRAA that are most important to the Army are: go twice as far, twice as fast. We beat that drum all the time. The other part that General Baker highlighted is that if we get those two things right, there's no doubt in my mind that this transformational capability will live on for many decades to come."

But how to design a platform which has such inherent capabilities but one able to adopt to a world where autonomous systems will be increasingly significant?

The Army approach encompasses the following considerations: Focus on the platform; focus on the payloads; focus on the con-ops of the formation; focus on the training with the resultant capability to work in flexible combat clusters.

Based on the discussion of the panel at AUSA which focused on the future of vertical lift, one can identify how the Army is planning to deal with each of these aspects of the way ahead for the FLRAA.

FOCUS ON THE PLATFORM

The Army is leveraging decades of experience of the USMC working with tiltrotor aircraft. This gives them a significant leg up with its new tiltrotor aircraft compared to the situation facing the Marines who blazed the path pioneering use of this new technology when they first took it to Iraq in 2007. The Army is leveraging the USMC experience in terms of the impact on con-ops, training, maintenance, and the entire experience of the tiltrotor enterprise which they now are writing a new chapter.

As Colonel Jeffrey Poquette, project manager for the future long-range assault aircraft, noted: "I know there is a plan to familiarize Army rotorcraft pilots. We have the V-22. The V-22 is a tiltrotor, so I've already started meeting a couple of Army aviators who have experience in the V-22. I just hired an experimental test pilot who is qualified in the V-22, so we're starting to build up the familiarity with a technology that the Army has not really used before."

There are a number of design aspects to the new tiltrotor which will enhance maintainability and performance of the aircraft, but it should be remembered that the V-22 is a bigger aircraft and one which has advantages due to its size as well.

But one of the key design features of FLRAA being crafted to meet the challenge of working with autonomous systems and the correlated systems onboard the aircraft to manage them is the use of a digital backbone designed to facilitate rapid upgrades.

This encompasses the hardware – the computerization – as well as the MOSA software. As Poquette noted: "MOSA, like I said, is such a big deal; I can't not mention it several times today. The digital backbone that enables MOSA didn't exist on Valor, right? So that's a significant part."

He added: "And if I could add, as an integrator of aviation systems, I work with counterparts in other PEOs whose job is to provide me the systems. We can't discount the importance of MOSA to integrating ASC quickly.

"Keeping pace with emerging threats is vital. ASC is one of the harder things to integrate with an aircraft. MOSA and the ability to leverage the digital backbone, leverage the standards that are open and available to the industry. Those standards, which are government-owned, are heavily informed by the architecture working group.

"I envision a future where as soon as a threat emerges, the necessary survivability equipment is ready for integration. The challenge lies not just in developing this technology but in ensuring it works effectively on the aircraft, which is where MOSA will take those timelines down a tremendous amount which is an important part of enhancing aircraft survivability."

The aircraft is being built with a digital engineering approach which allows as well significant ways to enhance design to production to upgrade capabilities as well. Given the focus on working the manned-autonomous systems effort in a combined arms approach the core point is really that the aircraft has a digital backbone focused on rapid upgrades through the use of the Modular Open Systems Approach or MOSA software as a gateway to rapid upgrades in payloads and systems.[3]

FOCUS ON THE PAYLOADS

Acquiring and integrating into the force of autonomous systems is very different from acquiring manned platforms. They are payloads carried by an autonomous vehicle (ground, air or maritime) and are focused primarily on a singular mission purpose or payload.

As Marcus Hellyer, the noted Australian strategist, put it in a recent interview I did with him in Canberra, Australia: "You can't look at autonomous systems as simply an unmanned version of a traditional platform. Everyone says that but I don't think they really think through what that means. And what it does mean is you don't want it to do everything that a traditional platform does with the autonomous system, because if you try and design it to do so, it's going to be just as complex as a manned system. This in turn means that is going to take just as long to design and it's going to cost just as much as a manned platform. There will be no savings in terms of time, money and people.

"In other words, the key point to underscore is this: Start simple, design autonomous systems to do one thing, and once they can do one thing effectively, and you work from there as the operators use them and input their demands into this process."

In the discussions by the panel of autonomous systems, it is apparent that they get Hellyer's points. For example, as Dr. Kirsch, Director of Combat Capabilities, underscored: "Our sister centers focus on some of the payloads, the different sensors, electronic warfare effects, and lethal effects. Our primary focus has been on behaviors and specifically how to get these launched effects to collaborate to accomplish our mission.

"We've done a lot of work showing how we can use launched effects with similar sensors for search areas or reconnaissance. In the next month, we're planning a capstone demonstration using a team-of-teams approach, where we have one operator managing multiple unmanned systems or launched effects with different capabilities. Some might have electro-optical/infrared sensors, some might be decoys, and some might have lethal effects. The operator will assign missions and decide when one sensor sees something and another needs to verify it. Ultimately, they'll pass that information back to a single operator, who will decide whether to prosecute the target or not.

"This involves a lot of collaborative behavior between the different platforms, allowing the operator to focus on their essential tasks while the autonomy of the launched effects manages the complexities of dividing the problem."

FOCUS ON THE CON-OPS OF THE FORMATION

Several speakers on the panel emphasized that the new manned platform would be working with "launched effects" which refers to either autonomous systems or loitering munitions to create the effects which an Army formation would be oriented to create.

As Major General Michael McCurry, Chief of Staff U.S. Army Futures Command. added: "From a tactical and operational perspective, certain mission sets align more naturally with autonomy. If we're putting 100 Rangers on the objective, there's likely someone flying that platform. Conversely, for repetitive tasks like logistics, where operations need to occur over extended periods, those are areas where we could see more rapid applications of autonomy.

"There are still considerations as we define behaviors. There's a natural alignment of certain mission sets. The other thing I would say about autonomy is that our focus is on protecting the humans. I talked about not sacrificing humans for first contact. Imagine a young Warrant Officer in the front seat of an AH-64 Apache with all that data coming at them. We can use different levels of autonomy to offload and untask saturate some of that workload, better protecting them and prosecute the mission more efficiently at the same time."

Or as Colonel Jeffrey Poquette commented: "And so, I would say from an S&T perspective, when you take into account the things that General McCurry was talking about — particularly in terms of the congested and contested airspace — as we look at launch effects, the sensors for launch effects, and autonomy, we're trying to use those capabilities to provide better situational awareness. This way, we can perform sense-and-avoid maneuvers so that we don't run into each other in that contested airspace.

"As we push our launched effects further out in front and conduct that scout role that, in previous years, we handled with manned assets, we need to understand what behaviors are required and what sensors are necessary to provide the same kind of awareness of what's out in front of us. So that, as we move manned formations forward, we know what we're about to face."

Army Brigadier General Cain Baker, Director of Future Vertical Lift Cross-Functional Team, highlighted the importance of the con-ops for formations as the shaping function for the kind of combined arms operations that FLRAA will do with autonomous systems for the future force:

"Generally, we're taking steps with this on the launched effects side for the Army. Last year, when we updated the Concept Document (CD), we went with both aerial launch capability and ground launch capability. So we're working through that now to determine the right mix of capabilities—whether launched at the formation level or employed from the air. A lot of that involves ongoing studies. We have an active study looking at that mix, and we're also incorporating simulations. We've always had high-fidelity simulations, but now we're actually putting it into our formation exercises. Additionally, we're also testing in the field. We just finished Edge, and I can discuss that if you like.

"Bringing formations out to Edge and then subsequently to Project Convergence helps us inform the DOTMLPF (Doctrine, Organization, Training, Materiel, Leadership and Education, Personnel, and Facilities) requirements. In the next year, we're also working closely with David's team and PM UAS on user demonstrations with launched effects inside formations, which will inform what type of formations we need and what employment capabilities they require.

"This speaks to a broader portion that General McCurry mentioned earlier about the system-of-systems approach. Launched effects interact with our ground forces, our aerial assets, and our Human-Machine Interface (HMI) on the ground, so there's a lot of learning going on here, and we're excited to get this out to the field."

FOCUS ON THE TRAINING

The speakers discussed the importance of training. From the standpoint of operational safety, training air crews is crucial. The experience with the Osprey underscores the importance of pilot training as 90% of crashes have been caused by pilot error.

Major General Michael McCurry emphasized: "No loss in training is acceptable. Every loss of a soldier is tragic, and when that happens, we do a deep investigation to determine root causes and implement corrective actions.

"In 2023, during my time overseeing aviation, we conducted an aviation stand-down focused on a bottom-up feedback exercise where we had units provide feedback all the way up. We collected it and backbriefed the Vice Chief and the Chief at the time, then went out to brief all the division and corps commanders in one session on what we found.

"That's where we discovered that, as we transitioned from the way we had been flying in Iraq and Afghanistan — higher above the terrain, with not a lot of threat to consider — to a more high-risk environment, we were overdriving our capabilities a little bit. Aviators were pushing themselves into situations they weren't yet prepared for...

"So, in aviation, we had to back up in 2023 and focus on some fundamentals, telling people it's okay to say, 'I need more training before I take that next step.' We took action to reorient the Department of Evaluation and Standardization at Fort Novosel to assist the CABs with training shortfalls. We simplified some maneuvers, particularly in the survivability realm, and focused on retaining those mid-grade warrant officers over time..."

But training also refers to learning how to operate in formations or what I call combat clusters in which combined arms operations are conducted with manned and autonomous systems.

Major General Michael McCurry put it this way: "When we're working on unmanned systems and launched effects, there are really three interrelated time periods. You've heard a lot about transforming contact, and General Schlosser mentioned earlier that soldiers figure out how to use things differently than engineers and designers intended. Transforming contact means we're giving some of these systems to engage soldiers so they can begin experimenting with them, informing organizational concepts, future doctrine, and training in the near term.

"These periods are interrelated: transforming contact informs the traditional deliberate transformation period and the program objective memorandum (POM). All of that is also being informed by the concept work that the Capability Command is doing. These three interrelated time periods — transforming contact, deliberate transformation, and concept-informed transformation — help us tie the picture together in a space that's moving quickly."

In short, one cannot understand the process of developing the Army's new tiltrotor aircraft into the force without understanding the impact of the age of autonomous systems into which it will enter the force, one which will do combined arms operations with synergy between manned and autonomous systems.

17

PUTTING THE OSPREY SAFETY RECORD IN PERSPECTIVE

T
he question of Osprey safety has been brought to the fore again with the accident in 2023 involving an Air Force variant of the aircraft.

The Department of Defense announced a return to flight of the aircraft in early 2024. This is the March 8, 2024 press release from NAVAIR announcing the return to flight:

Effective March 8, 2024 at 7 a.m. EST, Naval Air Systems Command is issuing a flight clearance for the V-22 Osprey thereby lifting the grounding. This decision follows a meticulous and data-driven approach prioritizing the safety of our aircrews.

A U.S. Air Force investigation began following the tragic loss of eight Airmen during the November 29, 2023, mishap off Yakushima, Japan. Our thoughts and prayers are with the families of the fallen.

In response to the preliminary investigation indicating a materiel failure of a V-22 component, the V-22 grounding was initiated on December 6, 2023. The grounding provided time for a thorough review of the mishap and formulation of risk mitigation controls to assist with safely returning the V-22 to flight operations.

In concert with the ongoing investigation, NAVAIR has diligently worked with the USAF-led investigation to identify the materiel failure that led to the mishap. Close coordination among key senior leaders across the U.S. Navy, U.S. Marine Corps, and U.S. Air Force has been para-mount in formulating the comprehensive review and return to flight plan, and this collaboration will continue.

Maintenance and procedural changes have been implemented to address the materiel failure that allow for a safe return to flight. The U.S. Navy, U.S. Marine Corps, and U.S. Air Force will each execute their return to flight plans according to service specific guidelines.

NAVAIR remains committed to transparency and safety regarding all V-22 operations. The V-22 plays an integral role in supporting our Nation's defense and returning these vital assets to flight is critical to supporting our nation's interests. NAVAIR continuously monitors data and trends from all aircraft platforms, so service members are provided the safest, most reliable aircraft possible.

The safety of our pilots, aircrew and surrounding communities remains of paramount importance.[1]

The Osprey has brought unique and transformative capabilities to the joint force and is a key part of the future as well. The Marines certainly have found that not having the Osprey in their exercise in the Nordics as altering what they can do and what they can bring to the defense of the Nordic region, a region in the throes of a significant integration effort, crucial to European defense and to the defense of North America.

In a 2024 article by MajGen (Retired) Steve Busby entitled "Groupthink gives V-22 a bad rap", the author drew on his experience and argued the following: *The V-22 has long gotten a bad rap. As soon as reports of a fatal accident involving an Osprey off the coast of Japan hit the internet last fall, the critics pounced, and a chorus of uninformed skeptics began posting and commenting, all asking: "Why is the Osprey still flying?"*

Supporters of the V-22 are quick to point out that the data tells a completely different story. In fact, the Osprey is a modern marvel in terms of performance and capability, and its operational safety record is on a par with the most widely used conventional rotorcraft flying in the Department of Defense today.

Like all first-generation cutting-edge technologies, the introduction of the world's first tiltrotor aircraft was a learning experience for everyone involved.

During initial development, the program suffered several tragic accidents including the loss of 19 Marines during an operational test flight in 2000.

That accident, more than any other single occurrence, damaged the reputation of the program. And, in the decades since then, the V-22 has been subjected to an overwhelmingly negative barrage of public opinion.

But facts matter, and the data shows the 10-year average mishap rate for MV-22s is 3.43 per 100,000 flight hours. For context, that places the Osprey's mishap rate squarely in the middle of the other type/model/series aircraft currently flown by the U.S. Marine Corps.

Examined another way, in the 17 years since the aircraft was first introduced into operational service in 2007, there have been 14 loss-of-aircraft mishaps across all three services and one international partner that operate the aircraft—or .82 mishaps per year while flying over 500,000 flight hours.[2]

Busby was there when the 3rd Marine Air Wing transitioned from the CH-46 to the Osprey. In fact, he presided over the last squadron to be trained for the CH-46 in anticipation of becoming a V-22 squadron.

This is the photo from February 22, 2013, and from a posting entitled: "Goodbye CH-46."

He was quoted in this 2012 story where the Osprey flew at the Miramar Twilight Air Show held in October 2012.

The aircraft has proven its superior capability for Marine Corps operations in Afghanistan, and it soon will have the same impact in the Pacific for transport of troops and supplies in security operations or humanitarian relief, said Maj. Gen. (select) Steven Busby, commanding general of the 3rd Marine Aircraft Wing headquartered at Miramar.

The primary advantage is its enhanced speed and range, coupled with aerial refueling capability. Those attributes effectively allow the Corps to replace a helicopter fleet with airplanes while retaining the ability to operate off ships at sea or areas ashore lacking runways.

"In the end what it means is the infantryman in the back, or whoever it is, is out of harm's way faster than anything else on the planet. And that's important to us," Busby said.

In the Pacific, with "the ability of that airplane to deploy with the KC-130s that provide refueling, we now have an asset that can range the entire theater. Either on a ship or without the ship ... it's going to be a game changer because higher, farther, faster is reality with that airplane."[3]

Safety is a crucial concern; but the ability of tiltrotor aircraft to work in ways that save lives is crucial as well.

That point was made in one interview which Ed Timperlake and I did when we visited in 2014 MAWTS-1, the Marine Corps's key weapons instructor training center.

During our time at MAWTS-1, we had a chance to talk with Captain Justin "Lumbergh" Sing who represents the new generation of Osprey operators who have not transitioned from other platforms. He noted: "I have not flown any other fleet aircraft. I went through the flight school syllabus and straight to the MV-22 FRS. Captain Sing had just joined MAWTS and had been there only three days."

He had two tours at sea with the 26th MEU as part of VMM-266(REIN). The 26th MEU was involved in the Odyssey Dawn Operation, but Captain Sing was part of the split Osprey force and was serving in Afghanistan during that operation. Sing served under Col.

Romin Dasmalchi for his first tour and Lt. Col. Christopher Boniface for the second.

During his time in Afghanistan, the Marines were expanding the operational envelope for the Osprey. As he noted: "We started utilizing V22 aircraft for the named operations in a new area previously unoccupied by U.S. forces while I was there."

He described one mission in Afghanistan in which the Osprey landed Marines and then quickly came back to move them out of harm's way.

The quick turn-around capability of the Osprey is an important capability for the "devil dogs" coming out of the back of an Osprey.

Captain Singh noted: "Two Ospreys inserted troops to a particular landing zone, one on either side of a tree line. We departed and repositioned to a laager point about 15NM away. Fairly soon after, we were called back to move the Marines out of a suspected IED infested area. They could not safely cross the tree-lined ditch at night.

"The next day we found out that the Landing Zone (LZ) where we had conducted the insert had IEDs in it. We just happened to not land on any. That was our first operation after our unit had just arrived in Afghanistan."

Captain Sing highlighted the quick turnaround time, which the Osprey was able to provide to the troops on the ground.

"From the time they called for immediate re-embark when we were on deck at the laager point, to the time they were repositioned, which included us landing, them loading, and us hopping the tree line and landing again was probably less than 15 minutes."

Captain Sing highlighted the impact of speed in an emergency medical situation as well.

"We were onboard the ship and had a sailor with a gallbladder issue. It was about to rupture, and they needed to get him to a medical facility.

"We were just north of Somalia in the Horn of Africa, and the closest medical care facility was in Mombasa down in Kenya.

"This happened while a party was being held on the flight deck, with no flight ops schedule that day. We needed to get this guy to medical care.

"The deck crew cleared the front half of the boat and pulled the V22 out on spot within 45 minutes, and we were in the air 45 minutes later.

"We had to tank on the way, but we had him on deck in Mombasa, Kenya roughly 1,100NM away within 4+30 hours after takeoff."

When asked how the Osprey had advantages over rotorcraft in approaching LZs, the Captain highlighted the advantage of a lower audible signature.

"We can maintain an audible standoff for a little bit longer by staying in airplane mode up at altitude and only descending when approaching the objective area.

"It really reduces the enemy's ability to know we're coming."[4]

18

STRATEGIC REDESIGN, THE 3 NS AND THE OSPREY

The Osprey's capabilities allow the joint force to do things they can not do with any other aircraft. And as the joint force evolves to operate in the new conditions of global strategic competition, the Osprey contributes in significant ways to that evolution,

In this chapter, I will highlight one such example. You have heard of the USMC Force Design 2030 as an adaptation to the threat environment. But with an emergent multi-polar authoritarian world, an even larger strategic redesign of U.S. forces is warranted

The Osprey is important for the first; but is crucial for the second.

The United States and its allies are shaping a significant strategic redesign in the defense of Northern Europe coupled with the force projected from Norfolk and North Carolina.

In a way what is being built with the Nordic region in terms of defense cooperation and integration (from Finland to Iceland) is being reinforced by the ability of the Norfolk-based fleet and the North Carolina-based Marines to work together in shaping a new strategic capability.

The three N's (Nordics. Norfolk and North Carolina) are reshaping Northern European and North Atlantic defense where the region is working its own defense and the United States (and Canada and Britain) provide reinforced capabilities to the overall strategic deterrence of Russia.

So how did this come about?

It is a significant story with several key interactive components but one still in progress.

But it is instructive for the future of U.S. and allied strategic redesign for the common defense, and not just in the North Atlantic.

THE NORDIC DEFENSE STRATEGIC REDESIGN

From the Nordic side, the new defense strategic redesign has been the evolution since 2014.

The Nordic nations are enhancing their national defense capabilities and over the past decade they are working more closely together on common solutions with the entrance of Sweden and Finland into NATO.

But this is not about Nordic NATO leaning heavily on the United States and its overstretched resources: it is about their own investments and enhanced efforts to work together.

I have travelled to the region frequently since 2014 and have seen these nations shape much greater cooperation than before 2014.

One example of this cooperation prior to the NATO expansion was the cross-border training among Norway, Finland and Sweden.

When I visited Norway in 2018, I went to the base where the Norwegians anchored their part of the tri-lateral air combat training.

This is what I wrote after my visit: *During my visit to Bodø Airbase on April 25, 2018, I had a chance to discuss the cross border air training which Norway is doing with Finland and Sweden. Norway is a member of NATO; Finland and Sweden are not.*

And with Finland to make a decision about its future fighter, that decision will affect the capability, which the three nations can deliver for integrated regional defense as well.

The day I was there, I saw four F-16s take off from Bodø and fly south towards Ørland airbase to participate in an air defense exercise.

The day before this event, the Norwegians contacted the Swedes and invited them to send aircraft to the exercise, and they did so.

The day before is really the point.

Major Trond Ertsgaard, Senior Operational Planner and fighter pilot from the 132 Air Wing, provided an overview to the standup and the evolution of this significant working relationship.

The core point is that it is being done without a complicated day-to-day diplomatic effort.

This is a dramatic change from the 1990s, when the Swedes would not allow entering their airspace by the Norwegians or Finns without prior diplomatic approval.

As Major Ertsgaard put it: "In the 1970s, there was limited cooperation. We got to know each other, and our bases, to be able to divert in case of emergency or other contingencies. But there was no operational or tactical cooperation. The focus was on safety; not operational training."

By the 1990s, there was enhanced cooperation, but limited to a small set of flying issues, rather than operational training. As Major Ertsgaard noted: "But when the Swedes got the Gripen, this opened the aperture, as the plane was designed to be more easily integrated with NATO standards."

Then in the Fall of 2008, there was a meeting of the squadrons and wing commanders from the Finnish, Swedish and Norwegian airbases to discuss ways to develop cooperation among the squadrons operating from national bases. The discussion was rooted on the national air forces operating from their own bases and simply cooperating in shared combat air space.

This would mean that the normal costs of hosting an exercise would not be necessary, as each air force would return to its own operating base at the end of the engagement.

The CBT started between Sweden and Norway in 2009 and then the Finns joined in 2010. By 2011, Major Ertsgaard highlighted that "we were operating at a level of an event a week. And by 2012, we engaged in about 90 events at the CBT level."

That shaped a template, which allowed for cost effective and regular training and laid the foundation for then hosting a periodic two-week exercise where they could invite nations to participate in air defense exercise in the region. From 2015 on, the three air forces have shaped a regular training approach, which is very flexible and driven at the wing and squadron level.

Major Ertsgaard added that "We meet each November, and set the schedule for the next year, but in execution it is very, very flexible. It is about a bottom-up approach and initiative to generate the training regime."

The impact on Sweden and Finland has been significant in terms of learning NATO standards and having an enhanced capability to cooperate with the air forces of NATO nations.[1]

When I wrote my co-authored book with Murielle Delaporte published at the end of 2020 entitled *The Return of Direct Defense in Europe: Meeting the 21ˢᵗ Century Authoritarian Challenge*, we focused on how the Nordics were leading the way in a new type of integration in European defense.

This is what we concluded: *Nordic defense and security cooperation are part of a broader global trend in which clusters of states are working together to enhance their ability to enhance their defense and security against the return of Russia and the rise of China. Clusterization is the next phase whereby liberal democracies do more for themselves in their joint defense rather than simply relying on diplomatic globalization initiatives through organizations like the EU or NATO to do that for them.*

"Clusterization" is key to generating enhanced capabilities that can work interdependently with key allies outside of a regional cluster to

reinforce the capabilities in a realistic and effective way to deter core adversaries. In the case of the Nordics, clearly the United States is the key outside power, with Brexit Britain and those states within continental Europe which have capabilities which can show up effectively to bolster the underbelly of the Nordic region are the key players that can reinforce Nordic defense. But at its heart, the Nordics need to bolster their own capabilities as well to work more effectively with their offshore allies and their continental European partners.

But to be blunt: this requires looking more realistically at what the defense of the Nordic region means against the evolution of Russian policies, strategies, and capabilities rather than simply to assume that NATO as a multimember alliance will simply show up.[2]

Now with all four Nordic countries becoming part of NATO, this process is accelerating.

But the real point for the United States is not to lead this effort but to support this effort.

And with the strategic changes put in place by the standing up of 2nd Fleet in Norfolk, such an approach is not only possible but built into the DNA of the re-launch of the fleet.

THE 2ND FLEET STRATEGIC REDESIGN: PROJECTING POWER FROM NORFOLK

The Obama Administration abolished the 2nd Fleet due to its perception of the diminished Russian threat.

It was disbanded in 2011 and folded most of its personnel, warships and responsibilities into Fleet Forces Command.

In the Trump Administration, 2nd Fleet was reestablished.

But the architect of the strategic design of the fleet, Vice Admiral Lewis, had in mind a different approach to fleet design than the Navy had built during the Cold War.

The template of the fleet during the Cold War was not re-applied by Lewis and his staff but a new approach was crafted and was needed to deal with the new global situation in which

U.S. assets were stretched and allies could do more in their own defense.

It is hardly necessary to say that this is the only template which is viable for the United States in age of multi-polar authoritarianism but one in which fiscal stringencies and over-stretched forces cannot be asked to do what the United States once did when it was the global economic leader and was in a binary conflict with the Soviet Union.

Ed Timperlake and I had the opportunity to spend time with Vice Admiral Lewis and his commands curing our frequent visits to Norfolk in 2021.

We described in detail the strategic redesign of 2nd Fleet and the standing up of the only NATO command on U.S. soil in chapter eight of our 2022 book entitled, *A Maritime Kill Web Force in the Making: Deterrence and Warfighting in the 21st Century.*

As Vice. Adm. Lewis put it to us: "We had a charter to re-establish the fleet. Using the newly published national defense strategy and national security strategy as the prevailing guidance, we spent a good amount of time defining the problem. My team put together an offsite with the Naval Post-Graduate school to think about the way ahead, to take time to define the problem we were established to solve and determine how best to organize ourselves to solve those challenges. We used the Einstein approach: we spent 55 minutes of the hour defining the problem and five minutes in solving it.

"Similarly, we spent the first two and a half months of our three-month pre-launch period working to develop our mission statement along with the functions and tasks associated with those missions. From the beginning our focus was in developing an all-domain and all-function command. To date, we clearly have focused on the high-end warfighting, but in a way that we can encompass all aspects of warfare from seabed to space as well."

In a speech in early 2021 to DSI's Fifth Annual Joint Networks Conference, Vice. Adm. Lewis underscored how he viewed the central role of allied and joint integration in shaping a way ahead for the commands. "At C2F, we have integrated officers from multiple allied nations directly into the fleet staff. The U.S. Marines, reserve component officers, and foreign exchange officers officer as the vice commander of C2F. At JFCNF, an initial team of fewer than ten In that speech, Vice. Adm. Lewis highlighted the importance of interoperability and interchangeability in working fleet capabilities.

"Interoperability is defined as 'the ability to act together coherently and efficiently to achieve tactical, operational, and strategic objects,' often involving the ability to exchange information or services by means of electronic communications. We must then be integrated—the ability of forces to not only work toward a similar mission, but to do so as one unit. An example of this is the Mendez Nunez, who deployed as part of the Abraham Lincoln Carrier Strike Group in 2019. The final step in the spectrum of relationships is interchangeability. That is the ability to accomplish the mission, regardless of which nation is executing a particular role."[3]

The three commands under the 2nd Fleet Commander are of course 2nd Fleet, then Allied Joint Forces Command and the Combined Joint Operations from the Sea Center of Excellence (CJOS COE) which is the only NATO Centre of Excellence on U.S. soil.

This command from its standup provided a unique blend of U.S. and allied cooperation and has focused naturally on North Atlantic operations and has embraced the developments in Northern Europe and are deeply concerned with Russian naval activity against North America, which has accelerated with the war in Ukraine.

Working Command and Control is a key element for shaping an effective integrated but distributed force.

But simply doing this at the fleet level is hardly sufficient.

This must be done with the relevant joint force — and in the North Atlantic airpower is the joint level by the USAF or the USMC or certainly in the case of the U.S. Army with regard to missile defense—and with the engagement of the capabilities, efforts, and interests of our allies.

And this requires that the U.S. Navy does a much better job of integrating with allied navies and forces, something which has been prioritized from the standup of C2F.

Vice. Adm. Lewis put this clearly in his speech to DSI's Annual Joint Networks summit cited earlier.

According to Vice. Adm. Lewis: "No one nation can face today's security challenges alone. The Joint service, allies and partners are force multipliers. Serving together, studying together, and participating in exercises together only increases our combined operational readiness. However, these relationships require time, effort, and we cannot assume that because they exist today, they will exist tomorrow.

"We value these relationships, and it takes concerted effort to build and maintain them — a critical advantage we hold over our competitors.... Maintaining security and stability in the Atlantic is a responsibility shared amongst many in order to ensure the international waters where we all operate remain free and open. Rather, it is a shared responsibility to ensure we are making changes to the way we operate TODAY, versus waiting until after hostilities start. We cannot afford to learn those lessons the hard way."

2nd Fleet is focusing on the High North as well as the more traditional aspects of North Atlantic defense.

And in doing so, it is working on ways to support and integrate with Nordic defense integration as well.

The High North is a very difficult place to operate, and the logistics challenge is significant but the ability to integrate Navy and Marine Corps operations with the Nordic nations opens up new vistas for operational effectiveness and resilience.

THE THIRD N: THE NORTH CAROLINA-BASED MARINES

In considering the Chinese as the pacing threat, the role of the

Marines based in North Carolina have seen their air capabilities reduced and the priority shifted to the Pacific.

But in a world of multi-polar authoritarianism, the pacing threat is global.

But with the limitations on U.S. resources, the difficult shift from the land wars and the absolute need to shift how the United States supports allies capable of largely defending themselves means that a dramatic shift in strategic redesign in Washington's foreign and defense policy and the force underwriting that policy is crucial.

In my view, the Marines based in North Carolina – II MEF and 2nd MAW – are doing so in conjunction with their Navy brethren in Norfolk who collectively are working a new relationship with the evolving Nordic defense efforts.

U.S. Marine Corps Cpl. Hayden F. Smith, with Marine Medium Tiltrotor Squadron 365 (Reinforced), sits on the edge of the MV-22 Osprey ramp during Exercise Trident Juncture 18 near Vaernes Air Station, Norway, Oct. 28, 2018. Trident Juncture 18 is an opportunity for II Marine Expeditionary Force (II MEF) to test and enhance their warfighting capabilities at the large-scale level in a unique, austere environment alongside their NATO Allies' and partners'. (U.S. Marine Corps photo by Lance Cpl. Camila Melendez)

After a visit to 2nd MAW in 2021, I expressed this opportunity as follows: *The North Carolina-based Marines have equipment pre-positioned in Norway and exercise frequently with the Norwegians. And through the Cold War and beyond, those Marines have had the mission to show up to reinforce Norway in a crisis.*

But in an era where there is a stated desire to have greater Marine Corps integration with the Navy how might this change?

And in what ways?

The answer in part needs to be generated by the geography., the missions and the allies.

The geography sees the growing role of the High North, and the question of using land space for operations rests on what particular allies will value and permit in a pre-crisis situation up to a full-blown crisis situation.

If one looks at the geography, it is clear the impact which enhanced Nordic integratability can have on rethinking what the Marines might do to reinforce the air-sea battle, which is really where the U.S. Navy is going in its reset to be able to fight and prevail in the 4th Battle of the Atlantic.

Given the priority concern which the Navy has with regard to Murmansk and the Kola Peninsula, those allies best positioned to reinforce U.S. and allied efforts are crucial to the warfighting and deterrence effort.

This means that Iceland, the Kingdom of Denmark (Faroe Islands and Greenland), Norway, Sweden and Finland are the anchors for effective deterrence in the region and can clearly shape the outcome with regard to any Fourth Battle of the Atlantic.

What can the Marines bring from North Carolina which would make the most SIGNIFICANT impact?

If one looks at the impact of Nordic integratability and the importance of reaching back to the air-maritime force operating across the arc of the North Atlantic with North America as the force generator up against Russian force projection, then two key roles for the USMC can be readily seen.

The first is to be able to provide the anchor for American forces operating from Nordic territory as part of the 360-degree air-sea-land battle.

By deploying a MAGTF to operate with all four Nordic nations, the Marines could provide not only contribute to enhanced forced integrability across the Nordic region, but with evolving C2, ISR and strike capabilities could reach back to the joint and coalition force operating in the extended battlespace.

In other words, they could support Nordic air-sea-land integration and support a distributed integrated U.S. and non-Nordic allied force as well. And they could make best use of the equipment which the Marines already own and operate as they gradually add new capabilities over the decade.

The second key role is for the North Carolina-based Marines to become the U.S. center of excellence with regard to working in the region. They could work language skills, including Russian. They would acquire the appropriate Cold weather gear so that they can deal with the old Norwegian saying that there is no bad weather, only bad clothing. And they could regularly exercise in the region and learn how to work with states that are reintroducing conscription and clearly focused on how to use civil society as part of the overall defense effort.

And the approach to MAGTF integration shaped in a Nordic context provides a number of technological and skill set evolutions relevant to any global role for USMC forces.

In short, rather than being a task for the U.S. Navy to sort through how best to leverage what Marines can do, the Marines would be at the center of working a key capability which clearly enhances the lethality and survivability of the fleet. Even better: this generation of Marines would learn how to prepare for the high-end fight with people whose very existence rests on getting this right.

A MAGTF working a regular exercise regime across the Nordic region could function as an anchor for the joint force and its engagement to provide scalable forces in support of Nordic defense, with the Marines providing a key role in working on the ground with regard to Nordic integration itself.[4]

Since I wrote that piece, 2nd MAW leadership has started down this path.

I interviewed MajGen Benedict, the Commanding General

of 2nd MAW in May 2024 shortly before he relinquished command.

U.S. Marine Corps Maj. Gen. Scott F. Benedict, third from left, commanding general of 2nd Marine Aircraft Wing, speaks with NATO service members during Exercise Nordic Response 24 in Alta, Norway, March 6, 2024. Exercise Nordic Response 24 is designed to enhance military capabilities and allied cooperation in high-intensity warfighting scenarios under challenging arctic conditions, while providing U.S. Marines unique opportunities to train alongside NATO allies and partners. (U.S. Marine Corps photo by Lance Cpl. Christian Salazar)

This is what I wrote about my meeting with Benedict in a May 2024 article: *MajGen Benedict and the 2nd MAW have been deeply involved over the last two years building relationships in Scandinavia culminating in the recent Nordic Response Exercise which saw Sweden and Finland participating as full members of NATO. For the USMC, the entry of Sweden and Finland into the NATO alliance means a substantial change from the primary focus on bolstering Norway in a crisis to being able to work with all the Nordic forces in a crisis; a facet they demonstrated by exercising with all three countries during the event.*

This means as well that the USMC can take its Marine Air Ground Task Force integrated capabilities and embed themselves within the network of Nordic defense and support the U.S. fleet as it operates in defense of the region.

In other words, it can operate from the land, within a significant defense belt provided by the Nordics, to support the fleet. The air capabilities of the USMC – Ospreys, F-35s, F-18s, and CH-53Ks – can operate from the land to support the fleet or to operate from the fleet to support of land-air operations. This capability to do either is truly unique and what the USMC brings to the fight – a significant force operating from the land to the sea and from the sea to the land.

The extensive training and integration with all the Nordic countries is a very significant development and lessons learned by Marines in this key region in the defense of the North Atlantic can be applied to the Pacific as well.

MajGen Benedict provided a wide-ranging picture of Marine Corps activities in shaping this concept of operations and the key role of 2nd MAW working with 2nd and 6th Fleets in the region.

He started by underscoring how he looked at USMC-Naval integration.

"I went to a senior commander" course in Naples where we focused on maritime combined arms operations. It struck me that both the Navy and Marines almost solely focus on Marine capabilities being employed from the sea, but not so much on how we can come from the land to support the naval campaign."

"The opportunity to work with the Nordics as they continue to enhance defense integration clearly allows us to demonstrate and take advantage of that opportunity and to shape innovative ways to do so. And we did that in the Nordic Response 2024 exercise as well. There is a lot we can achieve in littoral operations without solely operating from an amphibious ship."

We then turned to his experience during the exercise working with the air chiefs of the Nordic forces. He underscored that as they were working their way ahead, the Marines and the American forces are working closely on shaping effective C2 across the coalition force to operate as integrated as possible.

One should note that 2nd MAW brought its first squadron of F-35s to the exercise and with the Norwegians already operating F-35s, with Denmark and Finland to follow along with F-35s from the UK coming off of their carrier, which they did in this exercise off of the Prince of Wales. The F-35s are very interoperable with one another and are very capable of operating at a higher level of integration. When one adds German and Polish F-35s to the force, the capability is a substantial one.

The Finns in particular are masters of distributed air operations on their soil and the Marines worked closely with them and will continue to do so. The progress in this domain since I last talked to pilots at 2nd MAW working with Finns is significant. When I spoke to pilots at 2nd MAW in an earlier visit in 2018, they indicated that the Finns were teaching them about DO. Now the Marines are clearly working hard on their own approach to DO and having an ally like Finland who has lived on the shadow of a big power for a long time makes them a key partner in evolving DO for the F-35 as well.[5]

Benedict noted that focus of Force Design 2030 was upon being able to operate as the "inside force." But he underscored that the Nordics were the "inside force" and the role of the USMC was to reinforce their capabilities across the Nordic region which their air and sea capabilities were crucial to be able to do so.

The Marines have moved from their classic Cold War role of arriving in Norway and pulling out equipment from storage facilities as part of the reinforcement of Norway to become increasingly part of an integrated Marine Corps-Navy team reinforcing the Nordics who are enhancing their capability to defend themselves.

WHAT ROLE CAN THE OSPREY PLAY IN THIS STRATEGIC REDESIGN?

With the arrival of the CMV-22B to Norfolk, they can join with the 2nd MAW Ospreys to provide a shared capability. The Osprey and its operation from the Fleet with the Navy's CMV-22Bs or from the land via the Marine Corps MV-22Bs helps solve a key problem which was identified when I interviewed a logistics officer in II MEF in 2021 in solving this problem provides insight into how the Osprey enables the strategic redesign.

As the Marines work with the U.S. Navy to reshape capabilities for the maritime fight, two key elements for successfully doing so are the right kind of C2 for distributed integrated operations and logistical capabilities to support such a force.

The logistics piece is not an afterthought, but a key enabler or disabler for mission success.

With a sea-based force the force afloat has significant capability built in for initial operations, but the challenge is with air and sea systems to be able to provide the right kind of support at the right time and at the right place.

Engaging in operations against a peer competitor means that the force needs to be able to operate end to end in terms of secure communications and logistics.

Ensuring an ability to operate from home ports or allied ports is part of the security challenge; finding ways to use air systems to move key combat assets to the various pieces on the operational chessboard in the Atlantic is crucial; and having well placed and well protected stockpiled supplies which can be moved to support the force is a key part of the overall logistics puzzle which needs to be solved.

LtCol Perry Smith recently retired from the USMC but I inter-viewed him earlier this year when he was the senior strategic mobility officer for II MEF. He and his team focus on the end-to-end supply to the force, through air, sea and ground movements to deploying or deployed forces.

As he noted in a discussion at Camp Lejeune in April 2021, the Marines work end to end transportation which means that "the embarkers at the units actually do all the preparation for their own equipment, do all the certifications, do all the load planning, and move their units out."

But when force mobilization occurs for the joint force, the Marines are competing with the other services for lift support, and in North Carolina this means that they are competing with 82nd Airborne Divi-sion "for the same ports and airfields."

The logistics piece has two key elements.

- *First, there is the ability to support the initial deployment of the force.*

- *And secondly, there is the challenge of sustaining the force going forward.*

For the Marines, the logistics piece comes in two parts, namely, support afloat and support ashore, so there is a "naval slice and a ground slice."

For operations in the Atlantic AOR, the Marines are working with the Navy as well as key allies to work the logistics supply chain in a dynamic combat situation. This means that they need not only to work closely with the U.S. Navy but to be able to work closely with the support structures of key NATO allies in the support of European operations, including in the High North.

The Trident Juncture 2018 exercise provided an opportunity to work closely with the Norwegians on finding more effective ways to work with their domestic transportation systems, including capabilities like Norwegian ferries, to move equipment and supplies into the operational areas.

As Lt Col Smith put it: "What I saw at Trident Juncture was their willingness to make this plan work because they have to. I think they depend on us in a time of need to be able to do reception staging, onward movement, and get to the point where we can back them up in a fight if we needed to."

And to do this requires shaping as seamless as possible a logistics supply line.

As CNO Richardson stood up the Second Fleet, a key focus was on incorporating the High North into the shaping of new defense capabilities. To do so from a USMC point of view is challenging because of limited logistical infrastructure and the clear need to rely on air systems with fairly long legs, which means the Osprey and the coming CH-53K.

There is also the challenge of the environment.

As LtCol Smith highlighted: "In the Pacific, you don't have the problems we have in the High North with sub-zero temperatures with 24 hours of sun in the summer and two hours of daylight in the winter."

The Norwegians are very competent in such conditions and the Marines have a lot to learn from them, and leveraging the kind of clothing, and telecoms equipment which they deploy with would make a great *deal of sense.*

As LtCol Smith put it: "How do we take advantage of the knowledge of our allies and leverage their capabilities for our forces to enhance our own survivability and lethality?"

The communication challenges are significant.

As you operate from sea, and work with an expeditionary base, linking the two is a challenge, which requires having an airborne capability to link the two. When looking at the North Atlantic arc from North Carolina to the Nordics, strategic mobility is delivered by a triad of airlift, sealift and pre-positioning.

- *Where best to pre-position?*
- *How best to protect those stockpiles? And how to move critical supplies to the point of need rapidly?*

Reworking the Marine Corps force to work more effectively with the U.S. Navy requires a reset of the logistics enterprise.[6]

In what I am calling the "3N" strategic redesign, the Nordics are working collectively together to enhance their ability to operate in strategic depth across their region, in addition to enhancing local or national defense capabilities. 2nd Fleet and the NATO command are working to shape more effective maritime reach and cover over the region reaching back to North America.

The Marines can project into the region, and through their innovations in distributed operations in concert with the Nordic nations can work through various combat nodes across the region.

Supplies can be pre-positioned across the region and flown by Ospreys from the land to the fleet or by the Navy's own Ospreys from the fleet to the land base.

The Osprey whether operating from the ship or the land becomes a key logistical connector in the operations that can link up Nordic land and air defense with maritime reach.

Because the Osprey can land virtually anywhere, and has

speed and range, the logistical reach from sea to land or land to sea becomes a key enabler for the evolving strategic redesign of the defense of the Nordics and the reach into North Atlantic defense.

There is no other logistic link which can work the distribution of supplies embedded in the Nordic region with the range and speed of the Osprey in order to connect the various nodes of the warfighting force across an integrated land and maritime extended battlespace.

❧ 19 ❧
IN LIEU OF A CONCLUSION

W hen I worked for SECAF, he decided to implement a day without space for the USAF. It was not a pleasant experience. And we have now experienced what it is like to operate without an Osprey across the services. This has simply meant that core missions have not been met. Full stop.

When I went to visit MAWTS-1 in May 2024, I talked with the outgoing commander of this key training command, and discussed the recent WTI course. Unfortunately, the Osprey was not available due to the grounding of the aircraft by the three services.

This is what I learned from that discussion:

Col Purcell started by underscoring that the grounding of the Ospreys by the services after the accident last year with an Air Force Osprey, created a challenge for them. Not having Ospreys – which frankly are a bedrock platform in the transformation of their concept of operations – caused a problem in the WTI. There were some missions they simply could not do, and shifted assets around to do missions which was not their primary mission focus....

One mission which has been identified and which MAWTS-1 has been training for is the TRAP mission associated with a maritime strike mission. The need to recover rapidly any personnel downed in a maritime assault mission is something the Osprey is uniquely positioned to do. Only you can't do it if it is not there. Fortunately, the ban on Osprey use was lifted in time for them to be able to use the Osprey in the maritime strike event within FINEX.

But it does not stop there.

If you want to deliver an engine to a large deck carrier for the F-35C and don't have an Osprey, well you are out of Schlitz.

Or if the USAF is tasked with what President Carter asked the military to do in Iran in 1979, how would that work out? I remember specifically talking with my former professor Dr. Brzezinski about that mission failure and how the Osprey would have led to a different outcome, at least in his view.

You have heard of mission creep: but what missions missing in action?

That is what happens when you ground the tiltrotor enterprise.

And as a reminder of what the tiltrotor enterprise brings to the Navy this is what Rear Adm. Joseph Kilkenny, USN (Ret.) noted in a letter to the editor for *The Wall Street Journal* published on 2 October 2024:

Having commanded a carrier strike group, I can attest that the importance of tilt-rotor Ospreys in the Indo-Pacific can't be overstated.

The area in which U.S. forces must operate is vast. We need range, capacity and speed to resupply widely distributed forces. Only the Osprey has the combination of speed and range, with a small landing footprint, to make possible distributed operations in the Indo-Pacific. It can operate to and from ships and shore, including unimproved landing areas.

Sometimes a weapons platform transforms entire concepts of operation. The Osprey is one of those unsung examples. [1]

AFTERWORD

Ed Timperlake and I had the chance to interview LtGen (Retired) McCorkle for our book dealing with MAWTS-1.

Given the dedication of the book, it makes sense to include that interview as well in this book.

MAWTS-1 AND TRAJECTORY VISION: TALKING WITH THE 5TH COMMANDER OF MAWTS-1

December 13, 2023

We all know what tunnel vision is. But what is necessary for successful adaptation of a military in a dynamic situation is to have trajectory vision.

In other words, the ability to adapt but to do so with an eye on realistic adaptation which can be driven by new systems or new con-ops.

We had the chance to talk with LtGen Fred McCorkle about his involvement and experience at MAWTS-1 and how that shaped trajectory vision for the USMC which has been clearly demonstrated in adopting the Osprey and the F-35B and changing their con-ops to reflect doing so.

It is not just about adding some new technology; it is about understanding how best to leverage it to get the desired combat effect against a reactive enemy.

In our discussion with him, he underscored how the MAWTS approach was established and evolved to shape realistic innovation but with an eye to the future. And those who have been involved in the WTI experience were able to embrace change but had a "pretend I am from Missouri" kind of experience: show me.

This has led to MAWTS-1 being what we have referred to as an incubator for change. Or another way of putting it, rather than being trained to exercise tunnel vision, Marines have been trained to have trajectory vision or warriors who can embrace change. But not briefing chart change; real change demonstrated through what Lt General Rudder has called the "physics of warfare."

According to McCorkle: *I was a young major when first introduced to MAWTS-1. I went to the command, the DCA at the time was LtGen Fitch. When he first came, he said he was going to shut down MAWTS. But by the time he left his post, he had become one of the biggest supporters of MAWTS.*

We asked him, what did he think changed the DCA's mind?

McCorckle told us that the fact that MAWTS trained for integrated MAGTF operations was really the key.

He argued: *Those who came to the WTIs witnessed the best MAGTF training being done in the USMC.*

He credited John Lehman, then Secretary of the Navy, and the early commanders of MAWTs for setting in motion the framework to be able to provide for such integrated training. He also noted that Lehman allowed senior MAWTS officers, including himself, to receive clearances to have access to black Navy programs. This allowed them to have the possibility of trajectory rather than tunnel vision as commanders.

At the time, McCorckle noted that even the DCA's were not read into programs that the Admirals were allowed to be read into. This clearly created a problem as the introduction of new air systems and capabilities is a driving force for change.

One narrative McCorckle relayed was concerning the importance of the coming of night vision goggles. Although the night vision googles of the day were not perfect, McCorckle pointed out to general officers when he would have them experience low flights at night the following: *Would you rather have 20:50 vision with the NVGs or fly with 20:400 vision. They got the point.*

We then asked him about the safety of flight challenge.

McCorckle told his about his approach.

As CO of MAWTS I would brief every class and put up on the board a drawing. Here is this box. We want you to operate at the edge but not to go over the edge. You can operate near the corners of the box. I'll go to the mat for you to defend you if you operate that way. But if you go ¼ of inch outside of that box, I will be the first to "Recommend Pulling Your Wings"!

McCorkle also underscored the significance of MAWTS in terms of its COs having generally gone on to take General Officer Command and to bring their MAWTS experience forward and then re-invigorating MAWTS by contributing from their command position as well.

We then asked him about his Osprey experience and how MAWTS had prepared him for it.

He started by recounting a meeting with LtGen Fitch.

One day Fitch came up to me and put his arm around me and congratulated me on flying 65 different aircraft. I thanked him and told I was very humbled by the experience. He then underscored that he had flown 132 different aircraft!

Well, the Osprey was an aircraft like no other. Something which McCorckle experienced on his first piloting of the aircraft. It was certainly not a helicopter, a problem which still affects helicopter pilots who come to the Osprey and do not unlearn their rotorcraft skills.

As LtGen Fred McCorkle underscored: *At the end of a runway, a CH-46 will top end at 140 knots. If you are in loaded down Cobra, you are lucky to get to 120 knots at the end of the runway. And when I first flew it, I was at 250 knots at the end of the runway taking off.*

The test individual with me in the plane said we are limited in the speed we can go and you are already over the approved limit. I said the aircraft is virtually in neutral and we can go a lot faster and went to 330 knots.

This is what one can call trajectory vision at work

ABOUT THE AUTHOR

Dr. Robbin F. Laird is a long-time analyst of global defence issues. He has worked in the U.S. government and several think tanks, including the Center for Naval Analyses and the Institute for defence Analyses.

He is a frequent op-ed contributor to the defence press, and he has written several books on international security issues.

He is the editor of two websites, *Second Line of defence* and *defence.info*.

He is a member of the Board of Contributors of *Breaking defence* and publishes there on a regular basis.

He is a research fellow with The Sir Richard Williams Foundation.

He is also based in Paris, France, and he regularly travels throughout Europe and conducts interviews with leading policymakers in the region.

THE COMPANION BOOK

A TILTROTOR PERSPECTIVE: EXPLORING THE EXPERIENCE

By Robbin Laird

To be published 15 September 2025

This book complements the previously published book, *The Tiltrotor Enterprise: From Iraq to the Future*. The arrival of the Osprey into the USMC operational inventory in 2007 began a journey whereby the USMC was able to drive significant change in their concepts of operations by creating a tiltrotor enterprise. And these changes were accelerated with the coming of the F-35 to their force.

The U.S. Navy has added their variant of the Osprey to address the challenge of contested logistics, a key aspect of enhancing the force in an era of great power competition.

The U.S. Air Force has used the Osprey to do special force applications impossible with other platforms.

And now the U.S. Army is setting in motion a new chapter in the enterprise with their acquisition of a new tiltrotor aircraft and doing so in an age where autonomous systems will be key elements of the combat force.

In this volume, interviews with the warfighters and industry are provided which enhance the argument made in the first volume. In addition, there are insightful essays for analysts and practitioners of the tiltrotor art.

The two volumes together form a more complete sense of the experience generated by the tiltrotor enterprise.

AIRPOWER AND MARITIME FORCE MODERNIZATION SERIES

We have published a series of books which address air power and maritime force modernization and will be adding to this series as our work on these subjects progresses.

ITALY AND THE F-35: SHAPING 21ST CENTURY COALITION-ENABLED AIRPOWER

This book highlights Italy's significant role in the F-35 Joint Strike Fighter program. The book details Italy's contributions to F-35 manufacturing and maintenance at the Cameri facility, highlighting its status as a key production and sustainment hub for European and allied partners. Furthermore, the book highlights the integration of the F-35B into the Italian Navy's ITS Cavour aircraft carrier and its implications for sea-based operations. Finally, the book discusses Italy's participation in multinational exercises, such as Pitch Black, showcasing its long-range deployment capabilities and commitment to international partnerships, and the development of advanced pilot training facilities.

The book is based on interviews in Italy, the United States and Australia conducted with Italian pilots and airpower leaders since 2013.

Each chapter is presented first in the original English and is then followed by a translation in Italian. The translation was a machine translation and can be considered only an approximate one but having a translation in Italian can help the Italian reader to better understand the English and can provide for a wider audience as well for the book.

Published March 3, 2025.

THE COMING OF MARITIME AUTONOMOUS SYSTEMS: EMPOWERING AND ENHANCING THE KILL WEB FORCE

This book addresses the coming of maritime autonomous systems to the U.S. Navy and allied fleets. They are obviously coming as well to adversarial forces as well. It is clear that maritime autonomous systems will become a key part of the modernization of the United States (U.S.) and allied security and combat forces. They will also become part of the adversarial forces and will need to be countered as well.

Even though maritime autonomous systems are already here and ready to be deployed. there are the vested interests of maintaining a legacy approach to maritime operations and the thinking associated with such an approach which impede the way ahead.

This book lays out some key developments which have already occurred regarding maritime autonomous systems and what interactive changes between the arrival of such systems and the evolution of concepts of operations might be anticipated.

One such change is the emergence of a new class of ships, which might be called mother ships, and that dynamic is discussed later in the book.

Maritime autonomous systems are not ends unto themselves, but capabilities which enhance the distributed maritime force and its ability to contribute to joint or coalition operations across the spectrum of warfare.

Ranging from weapons to C2 to ISR to logistical payloads, maritime autonomous systems can deliver expanded capabilities to a Navy usually measured in terms of numbers and tonnage of capital ships.

In short, it is a whole new entry point into the future which empowers the force and provides for enhanced capabilities. But to do so is not just about technological development; it is about evolution of concepts of operations and evolving C2 and ISR working relationships for the fleet.

As LtGen (Retired) Steve Rudder, the former MARFORPAC Commander noted in the forward to the book:

"Dr. Robbin Laird has been leading the reporting on Unmanned Systems and Kill Webs for many years and has been producing forward thinking pieces on the evolution of autonomy. At each achievement, whether it be Ukraine, TF-59 in the Arabian Gulf, or the Australian Defense Force, his articles and books have provided a window into the future dominance of autonomous maritime systems and the journey into the Kill Web."

"The reader of Robbin's book should walk away with a sense of how autonomous maritime systems are changing how we think about Naval Warfare."

Published March 5, 2024

MY FIFTH-GENERATION JOURNEY: 2004-2018

This is a revised version of the book first published in 2023 and includes a new conclusion looking back at these early years from the standpoint of 2024.

It is a book about the coming of the F-35. It tells the story of those in the military who worked to introduce the aircraft into the combat force. Visits throughout the world accompanied the standing up the aircraft and this book tells the story of those visits and the work of the warriors who made the F-35 global enterprise possible.

According to the author: "It is my personal journey observing the development and evolution of the aircraft from my time working with the man who coined the term fifth-generation aircraft, Michael W. Wynne, through my many visits to F-35 sites, interviews with pilots, maintainers, and U.S. and allied government officials who navigated through the incredibly negative press and government officials trying to kill the program to have delivered a unique capability in the history of combat aircraft."

"It is a personal journey and I take the reader to many of the places where I went to talk with the F-35 nation. But I did so not only in the United States but in the nations of key members of the F-35 global enterprise. My journey is unique and I tell it not because of a desire to be remembered but in playing a role of recorder of history of those members of the military of many nations who made this capability real despite the media and many government officials desires to see them fail."

"But failure would have meant that we would have had even less capability than we have now after 20 years of fighting in wars of "stability" which have brought us the opposite."

As LtGen George Trautman, USMC (Ret), Former USMC Deputy Commandant for Aviation, writes in the forward to the book: "Robbin's ability to capture the perspectives of the key players, from pilots to maintainers and logisticians, provides a comprehensive and insightful account of this revolutionary aircraft. Moreover, the book uncovers the geopolitical implications of Fifth-Generation warfighting capabilities. As nations seek to assert their dominance and secure their interests, the strategic implications of these technologies ripple across the global stage. Through a series of personal essays and skilled interviews with those who understand the aircraft, My Fifth-Generation Journey: 2004-2018 deftly navigates through this complex web, painting a vivid picture of how Fifth-Generation warfighting will shape the future geopolitical landscape."

The book includes a number of original photos shot by the author during his visits highlighting the roles of pioneers in the setting up of the F-35 in the services and abroad.

Originally published August 16, 2023.

USMC TRANSFORMATION
SERIES

THE U.S. MARINE CORPS TRANSFORMATION PATH:
PREPARING FOR THE HIGH-END FIGHT

The United States Marine Corps began its modern transformation path after the introduction of the Osprey in 2007. In a series of in-depth interviews with the United States Marines, this analysis highlights the transformation strategy that has made the USMC one of the most dynamic military forces in the world today. From the land wars to dealing with peer competitor threats and engagements, this book demonstrates how the Marines are navigating the strategic shift to craft innovative solutions for the return of Great Power competition.

Many coalition partners look to the USMC as a relevant benchmark for the kind of multi-domain operations which they can pursue. For many allies, their force structure approximates the size of the USMC, and they find the fit better than emulating the total force which the United States has built. It is also the case that the legacy force coming out of the land wars is not directly applicable in terms of its warfighting relevance to the approaches for combat with the peer competitors.

"Only time will tell how the Marine Corps navigates this

treacherous transformation journey, but it's not the equipment that will make the Corps successful on the future battlefield – it's the Marines."- Lt-Gen George Trautman, USMC (Ret).

Published March 31, 2022

MAWTS-1: AN INCUBATOR FOR MILITARY TRANSFORMATION

Training for military forces is in the throes of significant change. The threats are dynamic; there is always the reactive enemy; and technology fosters new ways to operate. Concepts of operations are evolving, most notably as U.S. and allied forces are focusing on force distribution to deal with the higher end threats authoritarian adversaries are fielding.

Dr. Robbin Laird and Ed Timperlake have visited the major training centers in the United States and several abroad as the state of the art of training is dynamically developing as well. In this book, they highlight their visits to a major training center, MAWTS-1 located at Marine Corps Air Station, Yuma. This is a truly multi-domain training center and has been from its inception.

They have visited together or separately several times since 2011 so have seen the introduction of new air systems into the USMC and have been able to talk with Marines as they shifted from preparing for the land wars to the higher end fight generated by authoritarian adversaries. They have then had the benefit of talking with several former commanders of MAWTS-1 to gain further understanding of how MAWTS-1 was established and has evolved throughout its history.

As Lt Gen George Trautman, former head of USMC Aviation comments in the forward to the book:

"In their new book, MAWTS-1 An Incubator for Military Transformation, Robbin Laird and Ed Timberlake have captured the essence of Marine Aviation Weapons and Tactics Squadron One (MAWTS-1) in a very creative way...

"This book provides an analysis of the role of MAWTS-1 in the evolution of the USMC. MAWTS-1 keeps Marine aviation moving forward. New concepts of operation and new weapon systems demand refined tactics and innovative methods of training. As the Marine Corps exploits the incredible capabilities of the V-22 Osprey, F-35 Lightning II, CH-53 King Stallion, the TPS-80 Ground/Air Task Oriented Radar (G/ATOR) and the Common Aviation Command and Control System (CAC2S), the MAWTS-1 staff, and the Weapons and Tactics Instructors (WTI's) they produce, are at the cutting edge of keeping the Fleet Marine Force ready and relevant.

"The book tracks the recent evolution of the squadron, and then provides a series of interviews with the pioneers and Commanding Officers who developed and fostered the MAWTS-1 concept, fought off the naysayer's resistance to change, and led Marine aviation into what it has become today - an essential element of the MAGTF dedicated to bringing about a revolution in next generation aviation capabilities.

"These retrospectives may be the most valuable part of the book because they showcase the impact a small cadre of individuals, disappointed in their assessment of post-Vietnam Marine aviation, can have. Their visionary ideas set a new course that grew from something innocuously called "Project 19" into the premier aviation training squadron in the world - MAWTS-1."

Published June 6, 2024

THE COMING OF THE CH-53K: A NEW CAPABILITY FOR THE DISTRIBUTED FORCE

This book describes the coming of the CH-53K Kilo to the USMC and to its first international customer, the Israeli Defence Force. It is based on extensive interviews with the persons involved in the development, testing, build, and maintenance of the new combat air system. For air system it is -- built by the digital thread development and manufacturing approach, the aircraft is designed with maintainability and fleet support in operations as a key focus of the program,

If it were called CH-55 instead of the CH-53K perhaps one would get the point that these are very different air platforms, with very different capabilities.

What they have in common, by deliberate design, is a similar logistical footprint, so that they could operate similarly off of amphibious ships or other ships in the fleet for that matter.

But the CH-53 is a mechanical aircraft, which most assuredly the CH-55 (aka as the CH-53K) is not.

In blunt terms, the CH-55 (aka as the CH-53K) is faster, carries more kit, can distribute its load to multiple locations without landing, is built as a digital aircraft from the ground up and can leverage its digital backbone for significant advancements in how it is maintained, how it operates in a task force, how it can be updated, and how it could work with unmanned systems or remotes.

These capabilities taken together create a very different lift platform than is the legacy CH-53E. In a strategic environment where force mobility is informing capabilities across the combat spectrum, it is hard to understate the value of a lift platform, notably one which can talk and operate digitally, in carving out new tactical capabilities with strategic impacts.

The lift side of the equation within a variety of environments can be stated succinctly. The King Stallion will lift 27,000 lbs. external payload, deliver it 110 nm to a high-hot zone, loiter, and return to the ship with fuel to spare. What that means is JLTV's (22,600-lb.), up-armored HMMWV, and other heavier tactical cargos go to shore by air, rather than by LCAC or other slower sea lift means. For less severe ambient conditions or shorter distances than this primary mission, the 53K can carry up to 36,000 lbs.

With ever increasing lift requirements and advancing threats in the battlefield, there is no other vertical lift aircraft available that meets emerging heavy lift needs. There are a lot of platforms that can blow things up or kill people, but for heavy lift, the CH-53K is the only option.

The digital piece is a foundational element and why it is probably better thought of as a CH-55. This starts with the fly-by-wire flight controls. The CH-53K is the first and only heavy lift fly-by-wire helicopter.

The CH-53K's fly-by-wire is a leap in technology from legacy mechanical flight control systems and keeps safety and survivability at the core of the Kilo's design while providing a portal to an optionally piloted capability and autonomy.

The CH-53K's fly-by-wire design drastically reduces pilot workload and minimizes exposure to threats or danger, particularly during complex missions or challenging aircraft maneuvers like low light level externals in a degraded visual environment allowing the pilot to manage and lead the mission vice focusing on physically controlling the aircraft.

What this means is that the CH-53K "can operate and fight on the digital battlefield."

And because the flight crew are enabled by the digital systems onboard, they can focus on the mission rather than focusing primarily on the mechanics of flying the aircraft. This will be crucial as the Marines shift to using unmanned systems more broadly than they do now.

Published June 20, 2023

THE ROLE OF THE OSPREY IN THE PIVOT TO THE PACIFIC

The Osprey provides an important stimulant for the shift in con-ops whereby the Navy's experimentation with distributed operations intersects with the U.S. Air Force's approach to agile combat employment and the Marine Corps' renewed interest in Expeditionary Advanced Base Operations (EABO).

In other words, the reshaping of joint and coalition maritime combat operations is underway which focuses upon distributed task forces capable of delivering enhanced lethality and survivability.

The U. S. Navy's deployed fleet — seen as the mobile sea bases they are — faces a significantly different future as part of a distributed joint force capable of shaping a congruent strike capability for enhanced lethality. This means not only does the fleet need to operate differently in terms of its own distributed operations, but also as part of modular task forces that include air and ground elements in providing for the offensive-defensive enterprise which can hold adversaries at risk and prevail in conflict.

But how did we get here in 2023? How has the strategic shift for the joint forces evolved and caught up with what the tiltrotor revolution has enabled? And how has the Osprey evolved since the recognition of great power competition by the Trump Administration in 2018?

It began as a pivot to the Pacific in 2013; it is becoming a con-ops revolution enable in part by tiltrotor aircraft. The book takes two snapshots of this transition.

The first focuses on the introduction of the Osprey into the Pacific when the Obama Administration announced its "Pivot to the Pacific.

The second focuses on changes to the tiltrotor enterprise since 2019 after the Trump Administration highlighted the "Great Power" competition.

Published July 2, 2023

MAWTS-1: 2023 VISIT AND INTERVIEWS

This report focuses on MAWTS-1 in 2023. In 2023, I interviewed the CO of MAWTS-1, Col Eric Purcell, in April and then visited the command in November after the second WTI of the year. This provided a chance to discuss how MAWTS-1 had progressed in working enhanced force mobility for the USMC within the broader joint force, a key emphasis of the force design effort.

The challenge is that while the Marines are working FARPs and other means to enhance force mobility, the joint force is in the throes of significant change, whether it be the U.S. Navy working distributed maritime operations or the USAF working agile combat employment.

How does the USMC effort to reorganize and enhance its contribution to the joint force while the joint force is itself in fundamental change with much uncertainty over how to do maritime distributed operations and the agile combat air combat employment?

The Navy and Air Force sides of this transition have been a major part of our work published elsewhere and provide insights with regard to how challenging the overall force transformation is within which the USMC is working to find its proper place. It is not just up to MAWTS-1 to work the training for such an effort, but NAWDC and Nellis are clearly involved as well.

To put it simply: it is a work in progress and the Marines emphasis on a MAGTF organizing principle remains important going forward in spite of the effort to find ways to operate from much smaller organizational formations.

This report includes the interviews conducted in 2023. The date indicates when the interview was published on Second Line of Defense or Defense.info and collectively they provide an overview of how MAWTS-1 is training for the way ahead for the USMC by preparing the force that might have to fight tonite.

As the end of course video for WTI-1-24 starts: "It is not a question of if the Marine Corps will go into combat. It is only a matter of when."

Published December 5, 2023

THREE DIMENSIONAL WARRIORS: THE ROLES OF THE OSPREY AND THE F-35B

The publication provides an overview of the USMC will use the Joint Strike Fighter or F-35B and is using the Osprey as part of its operations. This is one of the first books which addresses how the USMC will use the Joint Strike Fighter and the Osprey to transform their operations.

The publication provides an overview of how aviation for the United States Marine Corps (USMC) shapes their entire capabilities and directs how they operate. USMC Aviation allows the USMC to be "three dimensional warriors" in fighting the "three block war." Not only docs aviation provide for 360-degree situational awareness, but aviation leverages the ground warrior against the "hybrid" enemies he faces today. As a former Commandant of the USMC characterized the challenge, the Marines have to be prepared to fight the "three block war." For then Commandant General Krulak, Marines needed to be prepared to operate over the spectrum of warfare within confined space. USMC aviation provides the essential glue for such capability.

The new elements being added to the USMC – the V-22 and the F-35B – provide a significant advancement in capability to support these concepts of operations. The role of the air element for the USMC is essential to its future. One can have a police force that wears military uniforms or one can have a flexible military force enabled by full spectrum capability. The choice depends upon the central role provided by an integrated air element for USMC operations and options. The air element is the strategic glue, which enables diversified, decentralized, and flexible USMC operations.

Published August 11, 2010

NOTES

INTRODUCTION

1. For our book on the kill web, see Robbin Laird and Ed Timperlake, *A Maritime Kill Web in the Making: Deterrence and Warfighting in the 21st Century*, 2022.
2. Robbin Laird, "2nd Marine Air Wing's 74th Birthday: Remembering by Building 21st Century Capabilities," *Second Line of Defense* (July 17, 2015), https://sldinfo.com/2015/07/2nd-marine-air-wings-74th-birthday-remembering-by-building-21st-century-capabilities/
3. Robbin Laird and Ed Timperlake, *MAWTS-1: An Incubator for Military Transformation.* 2024.

1. LOOKING BACK AND LOOKING FORWARD WITH THE OSPREY

1. https://www.24thmeu.marines.mil/News/Article/Article/510578/ch-46-helicopters-hmm-263-join-mag-29-to-support-oif/
2. https://military-history.fandom.com/wiki/VMX-22

3. A TILTROTOR ENTERPRISE

1. Dan Gouré. The Thirty-Year Success Story of the V-22 Osprey Must Continue," *Real Clear Defense* (April 15, 2021).
2. "Flying into 40: V-22 Program Office Recognizes Four Decades of Collaboration." *NAVAIR* (December 21, 2022), https://www.navair.navy.mil/news/Flying-40-V-22-program-office-recognizes-four-decades-collaboration/Wed-12212022-0835
3. https://www.navair.navy.mil/product/MV-22B-Osprey.
4. https://www.navair.navy.mil/product/CV-22B-Osprey.
5. https://www.navair.navy.mil/product/CMV-22B-Osprey.
6. Brian Everstine, "V-280's Speed, Range, Agility Right for FLRAA, Army Chief Says," *Aviation Week* (April 27, 2023), https://aviationweek.com/defense-space/aircraft-propulsion/v-280s-speed-range-agility-right-flraa-army-chief-says
7. "Soldiers conduct first touch point for long range assault aircraft," *U.S. Army* (December 26, 2023), https://www.army.mil/article/272697/soldiers_conduct_first_touch_point_for_long_range_assault_aircraft
8. https://www.dote.osd.mil/Portals/97/pub/reports/FY2022/army/2022flraai.pdf?ver=x9l6iuULMD91YWZ8mZu3uQ%3D%3D.
9. Meredith Rotten, "AUSA News: Army Wants Allies in on Future Vertical Lift," *National Defense* (October 10, 2022), https://www.nationalde

fensemagazine.org/articles/2022/10/10/army-partnering-more-closely-with-allies-on-future-vertical-lift

4. THE TIMELINE FOR THE TILTROTOR ENTERPRISE: A CON-OPS PERSPECTIVE

1. https://sldinfo.com/2009/11/general-walsh-on-the-usmc-use-of-airpower-in-iraq-from-precision-strike-to-presence/.

2. Robbin Laird, "2nd MAW Forward: The Role of Airpower in the Afghan Operation," *Second Line of Defense* (April 16, 2012), https://sldinfo.com/2012/04/2nd-maw-forward-the-role-of-airpower-in-the-afghan-operation/

3. https://sldinfo.com/2011/08/the-impact-of-events-avoided-the-key-role-of-the-agile-response-group-arg/.

4. "Pivot to the Pacific? The Obama Administration's 'Rebalancing' Toward Asia, *CRS* (March 28, 2012), https://sgp.fas.org/crs/natsec/R42448.pdf

5. Lance Cpl Benjamin Pryer, "MV-22 Ospreys Arrive at MCAS Iwakuni," *USMC* (July 23, 2012), https://www.mcasiwakuni.marines.mil/Iwakuni-News/News-Stories/News-Article-Display/Article/507246/mv-22-ospreys-arrive-at-mcas-iwakuni/

6. https://sldinfo.com/2013/08/the-osprey-and-innovation-breaking-the-mold/

7. *National Security Strategy of the United States of America* (December 2017), https://trumpwhitehouse.archives.gov/wp-content/uploads/2017/12/NSS-Final-12-18-2017-0905.pdf

8. https://www.2ndmardiv.marines.mil/News/Press-Releases/Article/2406055/ii-mef-conducts-regimental-air-assault-during-exercise-deep-water-20/

9. The two interviews are taken from Robbin Laird, *The U.S. Marine Corps Transformation Path: Preparing for the High-End Fight* (2021).

10. Robbin Laird, "The Deputy Commandant of Aviation Down Under: Plan Jericho Marine Corps Style," Second Line of Defense (March 18, 2016), https://sldinfo.com/2016/03/the-deputy-commandant-of-aviation-down-under-plan-jericho-marine-corps-style/

11. https://news.northropgrumman.com/news/releases/making-future-vertical-lift-open-safe-and-secure.

6. SEA LEGS: 2011-2015

1. https://www.militarynews.com/norfolk-navy-flagship/news/top_stories/bold-alligator-2011-revitalizing-amphibious-force/article_638a6a33-a8f0-5bb6-9036-7cd7b3e4603a.html

7. FOCUS ON PACIFIC OPERATIONS: THE PIVOT

1. https://sgp.fas.org/crs/natsec/R42448.pdf

9. THE HIGH-END FIGHT, CON-OPS AND FORCE DEVELOPMENT

1. https://www.forsvaret.no/en/exercise-and-operations/exercises/nato-exercise-2018
2. https://www.3rdmaw.marines.mil/Media-Room/Stories/News-Article-Display/Article/3256269/marine-air-control-group-38-refines-warfighting-capabilities/

11. AN OCTOBER 2024 UPDATE ON THE EAST COAST CMV-22B SQUADRON

1. https://www.airlant.usff.navy.mil/Press-Room/News-Stories/Article/3765300/vrm-40-welcomes-new-leadership-during-may-change-of-command-ceremony/
2. https://www.dvidshub.net/news/467954/first-east-coast-assigned-navy-cmv-22b-osprey-arrives-norfolk
3. https://www.airpac.navy.mil/Organization/Fleet-Logistics-Multi-Mission-Squadron-VRM-40/Leaders/Commanding-Officer/; https://www.airpac.navy.mil/Organization/Fleet-Logistics-Multi-Mission-Squadron-VRM-40/Leaders/Executive-Officer/

12. SHAPING THE TILTROTOR SUSTAINMENT ENTERPRISE: PHASES OF DEVELOPMENT

1. Robbin Laird, "The Osprey Maintenance Challenge: The View from 2010," Defense.info (February 11, 2019),https://defense.info/defense-systems/the-osprey-maintenance-challenge-the-view-from-2010/
2. https://www.dau.edu/acquipedia-article/performance-based-logistics-pbl-contracting-strategies
3. Robbin Laird, "Sustainment at the Point of Operational Relevance: Aligning the U.S. Navy's Sustainment Approach with Operations in the Contested Battlespace," *Second Line of Defense* (October 24, 2022), https://sldinfo.com/2022/10/sustainment-at-the-point-of-operational-relevance-aligning-the-u-s-navys-osprey-sustainment-approach-with-operations-in-the-contested-battlespace/
4. https://sldinfo.com/2024/06/making-a-good-aircraft-even-better-osprey-modernization/

13. THE OSPREY MAINTAINABILITY AND SUSTAINMENT ECO-SYSTEM

1. https://defense.info/defense-systems/ch-53k-logs-demo-a-shift-from-how-maintenance-was-launched-on-the-osprey/

14. SHAPING THE FUTURE OF THE ENTERPRISE

1. https://sldinfo.com/2010/11/an-update-on-the-osprey-from-new-river-4-major-york-on-expanding-the-battlespace-and-replacing-the-ch-46/
2. https://www.flightglobal.com/helicopters/army-special-mission-aviators-will-field-v-280-socom/158214.article
3. https://breakingdefense.com/2024/05/army-wraps-up-flraa-pdr-incorporating-special-ops-design-changes/
4. https://breakingdefense.com/2024/06/the-concept-driving-the-armys-air-assault-plans-for-the-indo-pacific/
5. https://breakingdefense.com/2024/06/the-concept-driving-the-armys-air-assault-plans-for-the-indo-pacific/
6. https://www.army.mil/article/278591/flraa_achieves_milestone_b

15. THE WAY AHEAD WITH FLRAA: INSIGHTS FROM AUSA 2024

1. https://www.janes.com/osint-insights/defence-news/air/ausa-2024-bell-helicopters-builds-first-flraa-parts-as-army-tests-tactics#:-:text=Bell%20Helicopters%20is%20finalising%20its,and%20development%20(EMD)%20phase
2. https://www.twz.com/air/armys-future-vertical-lift-tiltrotor-will-differ-significantly-from-v-280-valor-its-based-on#:-:text=The%20U.S.%20Army's%20Future%20Long,seats%20inside%20the%20main%20cabin
3. https://www.defenseone.com/business/2024/10/bell-presses-flraa-army-cools-large-programs/400328/?oref=defense_one_breaking_nl&utm_source=Sailthru&utm_medium=email&utm_campaign=Defense%20One%20Breaking%20News:%2010/16%20Audrey&utm_term=newsletter_d1_alert

16. LAUNCHING A NEW MANNED AIR SYSTEM AT THE DAWN OF AN AGE OF AUTONOMOUS SYSTEMS

1. https://www.defensenews.com/video/2024/10/16/ausa-future-vertical-lift/
2. https://www.army.mil/article/280035/edge_of_innovation_edge_24_concludes_at_u_s_army_yuma_proving_ground; https://www.defense.gov/Spotlights/Project-Convergence-Capstone-4/
3. https://breakingdefense.com/2024/10/mosa-a-winning-strategy-for-flraa-and-army-current-fleet-modernization/

17. PUTTING THE OSPREY SAFETY RECORD IN PERSPECTIVE

1. "NAVAIR Returns V-22 Osprey to Flight Status," *NAVAIR* (March 8, 2024), https://www.navair.navy.mil/news/NAVAIR-returns-V-22-Osprey-flight-status/Fri-03082024-0553

2. Steve Busby, "Groupthink Gives V-22 a Bad Rap," *Defense One* (February 25, 2024), https://www.defenseone.com/ideas/2024/02/groupthink-gives-v-22-bad-rap/394420/

3. Gretel C. Kovach, "Osprey in the spotlight at air show," San Diego Union Tribune (October 12, 2012), https://www.sandiegouniontribune.com/military/sdut-osprey-in-the-spotlight-at-air-show-2012oct12-htmlstory.html

4. Robbin Laird and Ed Timperlake, "Reflecting on the Tiltrotor Enabled Assault Force: The Perspective of a MAWTS-1 Osprey Instructor," *Second Line of Defense* (July 30, 2014), https://sldinfo.com/2014/07/reflecting-on-the-tiltrotor-enabled-assault-force-the-perspective-of-a-mawts-1-osprey-instructor/

18. STRATEGIC REDESIGN, THE 3 NS AND THE OSPREY

1. https://sldinfo.com/2018/07/the-nordics-rework-defense-the-role-of-cross-border-air-combat-training/

2. Robbin Laird and Murielle Delaporte, *The Return of Direct Defense in Europe: Meeting the 21st Century Authoritarian Challenge.*

3. Robbin Laird and Ed Timperlake, *A Maritime Kill Web Force in the Making: Deterrence and Warfighting in the 21st Century* (p. 273).Kindle Edition.

4. https://defense.info/re-shaping-defense-security/2022/02/the-usmc-and-embracing-the-dynamics-of-change-in-the-nordic-to-polish-defense-arc/

5. https://sldinfo.com/2024/05/majgen-benedict-on-his-time-as-cg-of-2nd-marine-air-wing/

6. https://defense.info/re-shaping-defense-security/2021/12/managing-co-evolution-of-the-us-navy-and-the-usmc-the-north-atlantic-case/

19. IN LIEU OF A CONCLUSION

1. https://www.wsj.com/opinion/its-a-bird-its-a-plane-no-its-a-v-22-osprey-774fe4b3?mod=Searchresults_pos20&page=1

Made in the USA
Middletown, DE
19 January 2026

27336359R00146